Autism Spectrum Disorder

Autism Spectrum Disorder

He Prefers to Play Alone

Todd T. Eckdahl

MOMENTUM PRESS
HEALTH

First published in 2018 by
Momentum Press®, LLC
222 East 46th Street, New York, NY 10017
www.momentumpress.net

ISBN-13: 978-1-94474-959-0 (paperback)
ISBN-13: 978-1-94474-960-6 (e-book)

Momentum Press Human Diseases and Conditions Collection

Cover and interior design by S4Carlisle Publishing Services Private Ltd., Chennai, India

First edition: 2018

10 9 8 7 6 5 4 3 2 1

Printed in the United States of America

Abstract

This book is an overview of autism spectrum disorder (ASD), an early childhood condition that affects 1 in 68 children in the United States, and an estimated 1 to 2 percent of children worldwide. ASD is characterized by impaired social skills, communication problems, and repetitive behaviors. The reader will learn about the early signs of ASD, and that the severity of ASD symptoms varies widely, ranging from mild effects with minimal impacts on functionality to severe effects that prevent independent living. The book explains that ASD is usually a complex trait disease caused by mutation of multiple ASD susceptibility genes, but that in rare cases it occurs as part of a monogenic genetic syndrome. The book describes genetic testing for ASD, and presents examples of ASD susceptibility genes that influence the development and function of the brain and the central nervous system. No treatments are available for the core symptoms of ASD, but the book describes therapies and drug treatments that can modulate them and address some of the health complications of ASD. The book presents new ASD biomarkers that can be used for better diagnosis, and explores the hopeful prospect of personalized medicine for ASD.

Keywords

Asperger's syndrome, autism, autism spectrum, autism spectrum disorder, complex trait disease, genetic disease, neurodevelopmental disorder, pervasive developmental disorder not otherwise specified

Contents

List of Figures

List of Tables

Acknowledgments

I am grateful to my friend Malcolm Campbell for encouraging me to take a leap of faith for this project, and for several others that have shaped my career as a science educator. I value Malcolm as a teaching and research collaborator, and I am proud of the positive impact that we have made together on science education and the improvement of science literacy. I am also grateful for the cheerful and professional support I received from the publishing team at Momentum Press.

This book would not have been possible without the support of my wife, Patty Eckdahl. She understands my passion for science and science education and helps me channel it in ways that benefit students and others around me. I also appreciate the support and encouragement that my parents, Tom and Bonnie Eckdahl, gave me in the pursuit of an education that would give me the privilege of sharing my love of DNA and genetics with undergraduate students and everyone else I meet.

I am grateful to my undergraduate genetics professor at the University of Minnesota, Duluth, Stephen Hedman, for helping me understand that I could pursue my love of genetics in graduate school. Thanks to John Anderson at Purdue University, who taught me to conduct molecular genetics research and to value undergraduate education. I appreciate the supportive environment that Missouri Western State University has provided me, and I am grateful to my mentors in the Missouri Western Biology Department, Rich Crumley, Bill Andresen, John Rushin, and Dave Ashley, who helped me learn how to engage students in the classroom and the research lab. I appreciate the many students whom I have worked with in class and collaborated with on research projects outside of class. I take pride in the contributions that my former students have already made, and will continue to make, to society.

Introduction

In 1935, a mother brought in her 3-and-a-half-year-old son for a psychiatric evaluation at Johns Hopkins University in Baltimore, Maryland. She explained that Alfred was a physically healthy child who had achieved physical developmental milestones such as feeding himself, gaining dexterity, and walking, but she was concerned about his psychological and emotional development. Alfred's mother believed that he was intelligent, and this was supported by the results of an IQ test, but she said that language had come slowly to him, and that after he learns a new word or phrase, he is content to repeat it over and over instead of learning new ones. The mother also had concerns about social interactions. Saying, "He prefers to play alone," she described that Alfred would climb down from playground equipment as soon as another child approaches, and that he paid more attention to objects than to people. At the age of 9 years, Alfred was again interviewed at Johns Hopkins. He was extremely tense, serious, and obsessively preoccupied with objects. Alfred spoke intelligibly during the interview, but he confused the meanings of words, often misused pronouns, and asked questions obsessively about topics of his own choosing, such as the lightness or darkness of the rooms in the clinic, and the letterhead on the patient history sheets.

The clinician who interviewed Alfred was Dr. Leo Kanner, a psychiatrist who took a special interest in children who were intelligent but had atypical social interactions, overwhelming desires to be alone, and extreme levels of routineness in their lives. Kanner detailed the case studies of Alfred and 10 other children in a 1943 paper entitled "Autistic Disturbances of Affective Contact," which marked a shift in thinking about **autism** and spurred research on its genetic, physiological, and psychological underpinnings that continues today. On the basis of interviews, diagnostic tests, and family histories, Kanner concluded that the commonly used descriptions of the children he interviewed as feeble-minded, retarded, moronic, idiotic, or schizophrenic were inappropriate and wrong. Instead, he offered the explanation that these patients had a previously unrecognized syndrome. He wrote that the children were born with an "extreme autistic aloneness that, whenever possible, disregards, ignores, and shuts out anything that comes to the child from the outside." This was consistent with descriptions of the children by their parents, such as "happiest when left alone," "acting as if people weren't there," "like in a shell," and "acting almost as if hypnotized." Although he borrowed the term autism from its use as a description of the introspective psychological state experienced by some people with schizophrenia, Kanner explained that the children he observed did not have schizophrenia and that their autism was not an alternative psychological state for them. Kanner later captured this concept with the term, **infantile autism**.

Following the publication of the seminal paper by Leo Kanner in 1943, gradual progress was made in the scientific understanding of autism, but beginning in the 1980s, there was a resurgence of research. Infantile autism was differentiated from schizophrenia in 1980 by the *Diagnostic and Statistical Manual of Mental Disorders* (*DSM*), a publication by the American Psychiatric Association that is widely used by clinicians, researchers, pharmaceutical companies, insurance companies, and lawmakers for the classification of mental disorders. The *DSM* listing of infantile autism was accompanied by diagnostic criteria that included onset during infancy, pervasive lack of responsiveness to others, impaired language development, atypical speech patterns, bizarre responses to environmental changes, and an excessive interest in objects. In 1987, the *DSM* replaced infantile autism with the term *autism disorder* and reorganized the list of diagnostic criteria into three categories: (1) qualitative impairment in reciprocal social interaction; (2) impairment in verbal and nonverbal communication and in imaginative activity; and (3) a restricted repertoire of activities and interests. In 1994, the fourth edition of the *DSM* (*DSM-IV*) introduced the term, *autism spectrum disorders*, to be inclusive of the specific diagnostic labels of autism disorder, **Asperger's syndrome**, and **pervasive developmental disorder not otherwise specified (PDD-NOS)**. Asperger's syndrome was first described in 1944 by Dr. Hans Asperger, a pediatrician who described four children who had trouble with social integration and nonverbal communication. In the *DSM-IV*, the symptoms of Asperger's syndrome are described as severely dysfunctional social interaction, impaired nonverbal communication, and restricted and repetitive patterns of behavior and interests. Spoken language acquisition often occurs at a typical pace in children with ASD, although unusual patterns of speech often develop, and these patterns were used to distinguish Asperger's syndrome from autistic disorder in the *DSM-IV*. PDD-NOS is described in the *DSM-IV* as causing severe and pervasive impairment of reciprocal social interaction or communication skills and stereotyped behaviors, interests, and activities. It was distinguished from autism disorder by occurring with an age of onset that is later than birth. The *DSM-5* was published in 2013, and it removed separate diagnostic labels for autism disorder, Asperger's syndrome, and PDD-NOS, referring to all of them as **autism spectrum disorder (ASD)**. The decision reflected the growing consensus among childhood psychiatrists and researchers that these three disorders have more similarities than differences and should be considered as part of a continuous spectrum. The *DSM-5* codified the diagnostic criteria for ASD as persistent deficits in social communication and social interaction, as well as restricted and repetitive patterns of behaviors, interests, or activities that arise in early childhood. The *DSM-5* acknowledged considerable variation in intellectual development and spoken language ability among people with ASD.

Changes in diagnostic criteria, better methods of ASD diagnosis, and increased awareness among both clinicians and the public are responsible for significant increases in estimates of the prevalence of ASD. In 2007, the Autism and Developmental Disabilities Monitoring Network (ADDM), funded by the U.S. Centers for Disease Control and Prevention, estimated the prevalence of ASD in the United States to be 1 in 150 children. The ADDM increased its estimate to 1 in 110 children in 2009, to 1 in 88 children in 2012, and to 1 in 68 children in 2014. The most recent estimate of the prevalence of ASD in the United States is similar to estimates of the worldwide prevalence of ASD, which range between 1 and 2 percent. Although ASD occurs in all ethnic groups, the ADDM reports that ASD is more common among Caucasian Americans than among African Americans or Asian Americans. The prevalence of ASD is also about 4.5 times higher among boys than girls. Improved understanding of ASD prevalence has brought increased attention to ASD, which has several benefits. Members of the public who might have family members with ASD are more aware of its signs and symptoms and can seek early diagnosis and early therapeutic intervention. Health care providers are more aware of diagnostic tools, therapies, and treatments for ASD, and scientists are better able to obtain research funding. Public policy makers are better able to support education of children with ASD, improve legal protections for children and adults with ASD, and support ASD research. Attention to ASD is also brought about by **World Autism Day**, established by the United Nations in 2007. On April 2, events are held throughout the world to promote awareness of ASD.

This book describes ASD as a developmental disorder that produces symptoms that impact functionality, but that vary widely in severity among individuals. Chapter 1 presents the signs and symptoms of ASD that occur in infants, children, and adults, and details methods for diagnosis. Evidence of the heritability of ASD, descriptions of genetic mechanisms that cause it, genetic testing, and contributing factors for ASD are explained in Chapter 2. Chapter 3 describes early interventions that benefit infants and children with ASD, therapies for ASD core symptoms and health complications, and pharmaceutical treatments for ASD symptoms. Prospects for improved ASD diagnosis, better drug treatments, and a personalized approach to ASD treatment are described in Chapter 4.

CHAPTER 1

Symptoms and Diagnosis

Describing autism as a spectrum expresses the view that the three developmental disorders of autism disorder, Asperger's syndrome, and pervasive developmental disorder not otherwise specified (PDD-NOS), are closely related, and produce a spectrum of overlapping symptoms. It also captures the observation that there is considerable variation in the primary symptoms and health complications of autism spectrum disorder (ASD) that produces a spectrum in the severity of functional impairment experienced by children and adults with ASD. The manifestations of verbal and nonverbal social communication deficits and restricted and repetitive patterns of behaviors, interests, or activities are different for each child with ASD, presenting unique challenges for early development and education. In adulthood, ASD brings challenges that affect the extent to which people can live independently and pursue productive and meaningful work.

Early Signs of ASD

Early signs of ASD occur in infants, but there are several reasons why they are difficult to notice. One is that there is considerable variation in the types and severity of the early symptoms of ASD among infants. The early signs are also subtle, and in many cases, they are not noticed at the time, but are recognized retrospectively after more obvious ASD symptoms arise. Early signs of ASD can occur in infants who do not have ASD, whereas some infants with ASD do not display some of the common early signs of it. Infants with ASD usually achieve physical developmental milestones such as rolling over, sitting up, crawling, and walking at a normal rate, and this makes it difficult for parents and healthcare providers to notice the subtle impact of ASD on the development of social and

Table 1.1 Early signs of ASD in infants and toddlers

Age	Signs of ASD
6 months	Does not make eye contact Does not return social smiles Does not express joy with smiles, laughter, and body movements
12 months	Does not babble or coo Does not use communicative gestures, such as pointing or waving goodbye Does not respond to name when called
16 months	Does not use words Displays intense interest in specific objects Repeats unusual body movements such as hand flapping and back arching
24 months	Does not use original short phrases Seeks sensory stimulation Prefers to play alone

communication skills. Table 1.1 lists the most common early signs that occur in infants with ASD. Infants with ASD do not return a smile or other expressions of joy as often as other infants do. Infants usually learn how to express happiness by the age of 6 months. Typically developing infants learn to make eye contact with others by the age of 6 months, but infants with ASD often do not learn this behavior at all, or make limited eye contact. These children do not often engage in joint attention, during which two people connect socially by looking back and forth at an object and each other. They have a deficit or delay in the development of reciprocal social interaction, during which emotions are shared with sounds and facial expressions. They do not imitate smiles or laughter as typically developing infants do, and they often fail to respond to the calling of their name. Sometimes they seem to completely tune out any sound from their environment, but at other times, they are oversensitive to faint sounds. Infants with ASD have deficits in nonverbal communication, which typically develop by the age of 9 months. Children with ASD frequently do not learn how to use gestures to communicate nonverbally with others, such as turning their head to draw attention to themselves or waving goodbye, as is typical by 12 months. They also appear to be less interested in interacting physically with others. Infants with ASD are less likely to seek cuddling, or to enjoy it. They often seem content to be alone, and often develop unusual habits of repetitive physical action, including rocking back and forth, hand flapping, and head banging. They have attractions to unusual objects, often preferring to carry around a hard object such as a metal car or a stapler instead of a soft one, such as a blanket or a stuffed animal. ASD also affects the development of verbal communication skills. Compared with typically developing infants, infants with ASD do not typically babble and coo within their first year, say their first word by the age of 16 months, or express meaning with two or more words by the age of 2 years.

People with ASD Have Impaired Social and Emotional Responsiveness

The early signs of ASD in infants and young children transition to symptoms in older children and adults. One symptom is an impaired ability to respond to others socially and emotionally, and it results in a variety of behaviors. People with ASD frequently have trouble beginning a social interaction in a way that is viewed by others as appropriate. They do not have a typical sense of personal space, and might start an interaction by touching someone inappropriately, or even licking them. People with ASD rarely initiate a conversation, except when they need help. It is uncommon for them to respond to direct communication, even when their name is used. During conversation, they struggle to maintain a dialogue. They frequently make statements out of context, and fail to clarify themselves if they are not understood. For example, one of the children described by Leo Kanner frequently interrupted their conversations by saying, "motor transports" and "piggyback," phrases with which she had become preoccupied.

Children and adults with ASD also have trouble developing shared interests with others. Despite having specific interests themselves, they often do not want to share them with other people. An obvious symptom of the impaired ability of people with ASD to respond to others emotionally is their failure to return a smile that is offered as a social greeting. It is rare for them to share their enjoyment or excitement, and they do not often display pleasure while interacting with others. People with ASD often show either an indifference to affectionate physical contact or reject such contact. They infrequently internalize the emotional states of others, and therefore do not offer comfort to them.

ASD Causes Deficits in Nonverbal Communication

Communication during social interaction involves both spoken language and nonverbal communication. Nonverbal communication in the form of facial expressions, gestures, and body language supplements the meaning of speech. Nonverbal communication is often difficult to control because it expresses underlying emotions. Many forms of nonverbal communication are instinctual, and cross the lines between cultures. They even cross the lines between species, indicating that they are primitive forms of communication that predate spoken language. A symptom of ASD is deficiency in the use of nonverbal communication during social interactions. Children and adults with ASD often fail to establish and maintain eye contact during a conversation, and fail to use body posture and position to engage in communication with another person, instead turning away from them during conversation. Leo Kanner observed this behavior in several of the children he interviewed, and reported that one of them never looked at the person talking to him and did not use gestures to communicate. When children and adults with ASD do use nonverbal communication, it is often an uncoordinated effort. For example, they might call attention to an object with hand gestures, but fail to reinforce the nonverbal communication by looking back and forth between the object and the person they are communicating with. People with ASD also have trouble integrating spoken language and nonverbal communication. They often fail to accompany their speech with appropriate facial expressions and gestures. Instead, they use gestures such as head nodding, head shaking, pointing, and waving in ways that are out of context with their speech, and do not contribute to their intended meaning. The ability of children and adults with ASD to convey meaning is often confounded by their use of atypical volume, pitch, intonation, stress, and rhythm during speech. People with ASD also have a deficit in their ability to understand and interpret nonverbal communication by others. The inability of people with ASD to use and understand nonverbal communication means they do not express their emotions clearly and have a limited ability to recognize and understand emotions conveyed by others.

People with ASD Have Difficulty with Relationships

Another symptom of ASD is having trouble relating to others in a variety of social situations. Children with ASD often have difficulties engaging in healthy play activities with other children. They seldom engage in cooperative play, during which children work together toward a common goal, and prefer parallel play instead, during which children play alone even though they are together. Although they engage in imaginative play, which includes role-playing and acting out past experiences, children with ASD rarely do so cooperatively with other children. They often fail to respond to social advances by other children, and are unaware that they are the subject of teasing. Children with ASD often pay more attention to objects than to other people, and seem unable to distinguish people from objects. For example, one of the children interviewed by Leo Kanner played with a hand presented to him as though it were a detached object, and blew out a lit match without any apparent acknowledgment of the person holding it. Another child responded to a slight pin prick with a reaction to the pin but not to the person holding it.

Children and adults with ASD often express emotions, ask questions, or make statements in an inappropriate social context. They often do not notice that another person is uninterested in an activity, or is upset about something. It is difficult for people with ASD to notice that they are being disparaged, or that they are unwelcome in a social situation. People with ASD also have a limited ability to recognize the effects of their behavior on the emotions of others. Many people with ASD have trouble making and maintaining close friendships with those who are not their primary caregiver. Some avoid friendships, whereas others are interested in making friends but lack the social awareness and skills to succeed in doing so. When children with ASD do make friends, their friends are often from a different age group. Some will display a partial or complete lack of interest in others. They often are unaware that others are present. When they do acknowledge others, they often show little interest in them, and have limited interaction with them. Many children and adults with ASD appear to be withdrawn into their own world. Leo Kanner learned of this type of behavior in one of the children he interviewed, whose father said that he displayed "an abstraction of mind which made him perfectly oblivious to everything around him."

ASD Causes Stereotyped Speech

ASD also causes children to adopt specific patterns of oral communication, referred to as **stereotyped speech**. Many children with ASD engage in **pedantic speech**, during which they use formal language to try to speak with authority and sound like an adult. They become attracted to specific jargon, but use it out of proper context and without meaning. Some children with ASD frequently produce unusual vocalizations such as squealing, humming, and squeaking. Children and adults with ASD often have **echolalia**, which is the frequent and often meaningless repetition of words or phrases spoken by others. Echolalia is related to the tendency by people with ASD to use rote language, during which they recite what they have memorized, such as the alphabet, an advertising jingle, or a famous poem. It is also related to the tendency to use **perseverative speech**, which is the repetition of a word or phrase after the stimulus that led to it has stopped. For example, a boy interviewed by Leo Kanner developed a habit of saying, "the people in the hotel," "candy is gone, candy is empty," and "don't throw the dog off the balcony." His parents explained that each of these phrases originated from a prior experience. People with ASD often use **idiosyncratic language**, during which they make reference to words and phrases that do not make sense without knowing where they came from. Kanner reported several examples of this type of speech among the children he interviewed. One child often uttered "chrysanthemum," "business," and "the right one is on, the left one is off." Another habitually said, "dinosaurs don't cry," "crayfish and forks live in children's tummies," and "gargoyles have milk bags." Children and adults with ASD often use **metaphorical language**, during which an object is used to represent an abstract idea. For example, when asked to subtract 4 from 10, a boy interview by Kanner responded, "I'll draw a hexagon." Another characteristic symptom of ASD is **pronoun reversal**. A typically developing child learns how to properly use personal pronouns by the age of about 3 years, but children with ASD often do not. They often refer to themselves in the second person instead of the first person, using pronouns such as "he/him/his," "she/her/hers," and "you/yours" instead of "I/me/mine." For example, one of the children interviewed by Kanner said "pull off your shoe" when he wanted his mother to take off his shoe, and "do you want a bath?" when he wanted a bath.

ASD Causes Restricted and Repetitive Behaviors

Children and adults with ASD develop abnormal **restricted and repetitive behaviors (RRBs)**. RRBs can take the form of ritualized patterns of behavior, such as turning around in a circle whenever a room is entered, or jumping up and down before every meal. RRBs also include repetitive hand movements such as clapping, finger flicking, flapping, or picking, stereotyped body movements such as swaying, spinning, rocking from side to side, and clasping hands to the ears. One of the children interviewed by Leo Kanner was known to walk about while repeatedly crossing his fingers in the air and shaking his head from side to side. Some people with ASD display intense body tensing, facial grimacing, and teeth grinding with an unusually high frequency and in situations that do not appear to others to be appropriate. Others display RRBs as routines associated with their daily activities. They might insist on having the same lunch every day, and always eat it in the same order, or on taking exactly the same route to school or the grocery store every time. They often develop rigid preferences for the foods they eat, the clothes they wear, and the products they use for personal hygiene or household cleaning. Those with ASD frequently resist changes that disrupt their routines and rituals, and get upset when even small changes require them to adapt their familiar behavior. The parents of one of the children Kanner interviewed reported that he "developed a mania for spinning blocks and pans and other round objects," and that when he was disrupted, he had violent temper tantrums. Maintaining a constant physical environment is important for many people with ASD, and they get upset when it changes. Because they often adopt routine and ritualized patterns of thinking, people with ASD find comfort in following rules, but have trouble thinking creatively.

RRBs can also take the form of ritualized patterns of verbal and nonverbal communication. Someone with ASD might adopt a repetitive line of questioning about a topic, and continue it long after they have received answers. They might insist on saying something in a particular, albeit unorthodox, way, and demand that others say it the same way. One of the children Leo Kanner interviewed insisted that, before every meal, his mother must say either, "Eat it or I won't give you tomatoes, but if you don't eat it I will give you tomatoes," or "Say 'If you drink to there, I'll laugh and I'll smile.'" People with ASD can find it difficult to understand the subtleties of humor. They tend to take language literally, and have trouble understanding statements that carry irony or implied meaning.

People with ASD often display RRBs because of very intense interests in a narrow range of concepts and objects. They often become obsessed with lists of things, such as numbers or letters, and are attracted to collections of information with simple and repetitive detail, such as a bus schedule, a dictionary, or a phone book. Examining lists of things is often a pleasurable activity for people with ASD, and they frequently enjoy making lists of their own. One child Leo Kanner interviewed made a list of all of the publication dates of *Time* magazine. Many people with ASD develop exaggerated attachments to collections of objects, such as trading cards, stamp collections, or old license plates. The restricted interests often take the form of a favorite object that they carry around. For a child with ASD, the object of choice is usually not the typical stuffed animal or blanket, but something unusual, such as a toilet plunger or a tennis racquet. An adult with ASD might develop an intense focus on a particular inanimate object, such as a pencil or a spoon, and derive pleasure from constantly having it with them and thinking about it. Frequently, children with ASD are fascinated with the parts of a toy, instead of the whole toy. They might, for example, play with just the wheels on a toy car, while other children push toy cars across the floor.

ASD Causes Atypical Sensory Responses

Symptoms of ASD are also associated with the response to sensory input and the level of interest in it. People with ASD sometimes have a smaller than expected reaction to sensations, but sometimes they have a larger one. For example, they might be less affected by extremes in temperature than people around them, but get upset when their hair is cut. Children with ASD might not cry when they experience pain that would make other children cry. Children and adults with ASD often develop a deep interest in the visual sensation they get from watching simple and repetitively moving objects, such as spinning wheels, ceiling fans, or rocking chairs. Many children and adults with ASD explore unusual sensory inputs, such as licking objects to see what they taste like, or sniffing them to discover new smells. Some develop sensitivities or phobias to some sounds, and are content to ignore others. For example, they might become highly sensitive to the sound of thunder, but do not react at all to a loud car horn. Leo Kanner described a child who developed a deep fear of the loud noises made by meat grinders, vacuum cleaners, and trains, but also became obsessively interested in them. People with ASD often pay more attention to the texture of things than most people do, and sometimes develop an aversion to being touched by certain objects.

Diagnosis of ASD

Individuals with ASD display considerable variation in the symptoms they develop, the age at which symptoms arise, and the severity of the symptoms. Because many of the symptoms of ASD can also be caused by other developmental disorders, ASD is very difficult to diagnose. ASD can be diagnosed by the age of 18 months, but most children are not diagnosed until past the age of 2 years, and some are not diagnosed until adolescence or adulthood. The American Academy of Pediatrics and the Centers for Disease Control and Prevention recommend ASD developmental screening during 18-month and 24-month well-child visits. Diagnosis of ASD in infants and toddlers is a two-step process consisting of developmental screening and a comprehensive diagnostic evaluation. For infants, screening and diagnosis begins with detection of at least some of the early signs of ASD (see Table 1.1). Informal observations during ASD screening include watching how infants make eye contact, respond to others, gesture, communicate nonverbally, and react to their spoken names. Family observations are useful, as well as family histories of ASD or other developmental disorders. Formal observations can be made with screening tools developed to assess the risk of ASD that have sensitivities and specificities exceeding 70 percent. The **Modified Checklist for Autism in Toddlers, Revised (M-CHAT-R)** is designed for toddlers between the ages of 16 and 30 months, and can be taken at a clinic during a well-child visit, or online (URL in Bibliography). The checklist asks parents or caregivers a series of 20 questions about the child and uses the answers to determine a low, medium, or high risk of ASD. The M-CHAT-R has a high false positive rate, and can produce a medium or high risk score for children who do not have ASD. Because of this error rate, the authors of M-CHAT-R developed a series of follow-up questions that ask caregivers to provide specific examples of the behaviors that can be used to make a more accurate assessment of the risk of ASD. The **Screening Tool for Autism in Toddlers and Young Children (STAT)** assesses the risk of ASD in children between the ages of 24 and 36 months, and takes about 20 minutes to administer. The STAT is an interactive test during which the examiner engages the child in communication and play, makes observations about the child's behavior, and determines a score that represents the risk level for ASD.

In many cases, the signs and symptoms of ASD are either not noticed during infancy or they develop later. Among the possible signs of ASD are avoiding eye contact, lack of empathy for the emotions of others, and preferring to be alone. Highly restricted interests, RRBs such as hand flapping, finger flicking, rocking, and spinning, having unusually extreme reactions to sensory inputs, and getting upset with minor changes in routine or physical environment can be ASD signs as well. Delays in language acquisition and unusual patterns of speech such as echolalia are also indicative of ASD. It is especially hard to diagnose ASD on both ends of the spectrum of severity. For people who are highly functioning and have good verbal communication, the behavioral symptoms of ASD are subtle and mixed with typical behaviors. For those who are nonverbal and have cognitive disability, it is challenging to distinguish ASD from other conditions.

A positive result from developmental screening for ASD of an infant or toddler justifies a comprehensive diagnostic evaluation by a specialist such as a child psychologist, a child psychiatrist, a pediatrician or a child neurologist. A critical part of the evaluation is making careful observations of the behavior and symptoms of the child in a clinical setting, in the home, or in a classroom. There are several diagnostic tools to standardize these observations. The **Gilliam Autism Rating Scale-Second Edition (GARS-2)** is for ages 2 to 22 years, and requires only 10 minutes to gather information about characteristic behaviors associated with ASD. The **Childhood Autism Rating Scale-2 (CARS2)** is for children aged 2 years and older, takes about 15 minutes to complete, and produces a score that reflects the severity of autistic behavior associated with relationships, verbal and nonverbal communication, and emotional responsiveness. The

Autism Diagnosis Interview-Revised (ADI-R) is a structured interview that takes about 2 hours to administer and score. It measures reciprocal social interaction, communication and language, as well as restricted and repetitive interests and behaviors shown by children aged 2 years and older. The **Autism Diagnostic Observation Schedule-Generic (ADOS-G)** assesses social interaction, communication, play, and imaginative use of materials. During four 30-minute sessions, an examiner observes the occurrence or nonoccurrence of behaviors that are associated with ASD.

Interviews with parents or other caregivers are also part of a comprehensive diagnostic evaluation for ASD. Topics include complications during pregnancy or birth, the achievement of physical developmental milestones, the occurrence of serious medical conditions, and the incidence of injuries. Of special interest are behaviors associated with eating, sleeping, and other daily activities, as well as the development of nonverbal and verbal communication skills. Standardized caregiver interviews are also used, such as the ADI-R, which queries reciprocal social interaction, communication and language, and behavioral patterns in children 18 months and older.

Because an important part of ASD diagnosis is ruling out other neurological causes of the observed symptoms, neurological tests are often conducted. For example, magnetic resonance imaging (MRI) is used to uncover structural brain abnormalities and electroencephalography (EEG) tests investigate alternate causes of seizures. Vision and hearing tests clarify the effects of possible hearing or vision problems that might contribute to symptoms associated with ASD.

A definitive diagnosis of ASD is important for initiating an appropriate course of treatment, therapy, and intervention. Children will benefit from appropriate educational services, and access to disability-related services and accommodations can help children and adults alike. A diagnosis of ASD is made by comparison of the results of a comprehensive diagnostic evaluation with criteria for ASD, such as those found in *Diagnostic and Statistical Manual of Mental Disorders*, Fifth Edition (*DSM-5*). Although the *DSM-5* has come under criticism for being subjective, lacking reliability, and containing cultural biases, it is the most widely used method in the United States to diagnose ASD. There are five *DSM-5* criteria for ASD. The first criterion concerns deficits in social communication and social interaction. It describes deficits in social–emotional reciprocity, in nonverbal social communication, and in developing, maintaining, and understanding relationships. Restricted and repetitive patterns of behaviors, interests, or activities are covered in the second criterion, including stereotyped or repetitive movements and speech, insistence on routines and rituals, highly restricted and intense interests, and atypical reactions to sensory inputs. The third *DSM-5* criterion for ASD is that the symptoms covered by the first two criteria must be present in early childhood. To accommodate the highly variable nature of ASD, the *DSM-5* recognizes that symptoms might not fully develop until social demands expose them, or they might be covered up by compensatory behaviors in older children and adults. The final two *DSM-5* criteria for ASD are that the symptoms cause clinically significant impairment of the ability of the patient to function in social or occupational settings, and that the observed symptoms cannot be better explained by an intellectual developmental disorder or a global developmental delay.

The *DSM-5* also includes descriptions of symptoms associated with social communication and RRBs that can be used to determine the severity of ASD. Level 3 is the most severe level and is characterized by severe deficits in verbal and nonverbal social communication that limit social interactions, and RRBs that interfere with function. People with level 3 ASD require very substantial support because they are often either nonverbal or cannot speak in a way that is intelligible to others. They rarely initiate social interactions, but when they do, they often make unusual social approaches only to meet their needs. Level 2 ASD is characterized by marked deficits in verbal and nonverbal social communication that impair social interactions, and by inflexibility of behavior that interferes with function. Children and adults with level 2 ASD require substantial support. They often speak in simple sentences, interact with other people with regard to narrow special interests, and have atypical nonverbal communication habits. Level 1 ASD occurs when there are moderate deficits in social communication that cause impairments in social interaction, and when there is inflexibility of behavior that results in significant interference with functioning. Children and adults with level 1 ASD require some support.

Although the *DSM-5* is the most popular method of classifying ASD in the United States, the official diagnostic manual for the country is the **International Classification of Diseases (ICD)**, which is used widely in Europe and the rest of the world. Maintained by the World Health Organization, the ICD is in its tenth edition, known as ICD-10. Although the *DSM-5* does not make a diagnostic distinction among autism, pervasive developmental disorder, and Asperger's syndrome, the ICD-10 does. The ICD-10 contains separate classifications for childhood autism, Asperger's syndrome, atypical autism, other pervasive developmental disorders, and unspecified developmental disorders. The difference between the diagnostic approaches taken by the *DSM-5* and the ICD-10 illustrates an ongoing debate over whether there really are distinctions to be made among these developmental disorders.

ASD and Intellectual Disability

About 1 to 3 percent of people worldwide have **intellectual disability**, which occurs when there are deficits in the ability to reason, think, and learn that limit the ability to lead a productive and independent life. The most common measure of intellectual capacity is the **intelligence quotient (IQ)**, which expresses the results of standardized tests. The most widely used IQ tests are the **Wechsler Adult Intelligence Scale**, for ages 16 years and older, and the **Wechsler Intelligence Scale for Children**, for children aged 6 to 16 years. The **median** IQ score for children or adults of a given age is set at 100, and the range of 85 to 115 includes all IQ scores within one standard deviation of the median. Because the range of 70 to 130 contains all the scores within two standard deviations, 96 percent of all IQ scores fall within this range. About 2 percent of people have an IQ below 70, and are considered to have an intellectual disability. Those with an IQ of less than 50 are considered to have moderate intellectual disability, whereas those with an IQ of less than 30 have severe intellectual disability. The causes of intellectual disability vary widely, and can be categorized as environmental or genetic. The most common genetic cause of intellectual disability is Down syndrome, which occurs at a rate of about 1 in 700 births, and is usually caused by the inheritance of an extra copy of chromosome 21. About 20 percent of intellectually disabled people have Down syndrome.

Accurate assessment of intellectual disability among children and adults with ASD is hindered by general challenges associated with measuring intelligence at all, but also by specific challenges that come from measuring it in in the context of ASD. Poor communication skills, deficits in social interaction, sensitivity to environmental stimuli, and stereotyped behaviors can interfere with performance on IQ tests, resulting in underestimates of intelligence. Data associated with intellectual disability among people with ASD should be interpreted with caution. One source of data is a 2010 study by the Centers for Disease Control and Prevention, which showed that 31 percent of children with ASD had intellectual disability, defined by an IQ score of 70 or below. The study found that an additional 23 percent of children with ASD had borderline intellectual disability with an IQ score in the range of 71 to 85. When assessing intellectual disability in children and adults with ASD, medical professionals supplement information from IQ tests with developmental tests. Two commonly used tests are the **Denver Developmental Screening Test**, which measures the ability of preschool children to carry out tasks in the four categories of social contact, fine motor skill, gross motor skill, and language, and the **Vineland Adaptive Behavior Scale**, which assesses the ability of children to cope with environmental changes, to learn new skills for daily living, and to display independence. The **Comprehensive Assessment of Spoken Language** and the **Clinical Evaluation of Language Fundamentals** measure spoken language expression and understanding, and are used to clarify the severity of intellectual disability in children with ASD.

Savant Syndrome and ASD

Some people with ASD and other developmental disorders have **savant syndrome**, which means that they have prodigious talents or intellectual abilities despite having deficits in other aspects of cognition that cause functional disability. Estimated to be about 1 in 100, the occurrence of savant syndrome among people with ASD is about 20 times higher than it is for people with other developmental disorders. Some people with savant syndrome have the ability to memorize large amounts of information, such as an entire phone book, a dictionary, or a collection of poems, whereas others display exceptional artistic talent or unusual musical abilities, such as perfect pitch or the ability to identify a piece of music from a single chord. Some people with savant syndrome have **hypercalculia**, and can very quickly add, subtract, multiply, and divide large numbers. Some of those with hypercalculia are **calendrical savants**, and can quickly calculate the day of the week for any date in history. Savant syndrome was on display in the 1988 movie, Rain Man, in which the main character, Charlie Babbitt, had unusual abilities for counting and memorizing, despite having functional disabilities that prevented him from pursuing a career and living on his own. The character was based on a real person, Kim Peek, who read and retained an estimated 98 percent of the information in thousands of books, making him a walking encyclopedia on a variety of subjects, including literature, history, geography, music, and sports. He could provide driving directions between almost any two cities in the world, and was a calendrical savant. Although Charlie Babbitt clearly displayed the symptoms of ASD, Kim Peek did not have ASD. He was born with a different developmental disorder caused by **agenesis of the corpus callosum**, a brain defect in which there was no communication between the left and right brain hemispheres.

ASD Health Complications and Comorbidities

A variety of health complications occurs in children and adults with ASD. Some stem from ASD itself, but others come from the presence of one or more other disorders, which are called **comorbidities**. Some ASD comorbidities are psychological and behavioral disorders that produce symptoms that contribute to cognitive and functional impairment. Most children with ASD have **anxiety disorder**, a comorbidity that causes them to maintain a state of fear about the future and a heightened stress response to perceived dangers. Mild anxiety takes the form of tenseness, apprehension, and shyness, and can cause physical symptoms such as sweating, headaches, and fatigue, whereas severity anxiety is associated with restlessness and irritability, and can cause tremors, breathing problems, hypertension, nausea, insomnia, and changes in appetite.

Approximately one-third of children with ASD develop **attention-deficit/ hyperactivity disorder (ADHD)**, which causes them to have trouble paying attention and focusing on tasks, and to act impulsively without thinking. Early signs of ADHD include frequent fidgeting and squirming, an inability to sit still, nonstop talking, and a habit of interrupting people. ADHD often persists into adolescence and adulthood, and can cause people to have poor organizational skills, to have trouble concentrating, to become forgetful, and to avoid tasks that require sustained and focused mental effort. Children with ASD are also more likely to have **developmental coordination disorder (DCD)**, a comorbidity in which inaccurate transmission of nerve signals from the brain to the body causes a delay in the development of motor skills and impairment of physical coordination. Sometimes referred to as clumsy child syndrome, DCD prevents children from learning basic skills such as eating with utensils, brushing their teeth, dressing themselves, and riding a bicycle, and school skills like using a pencil, a scissors, or a computer. The physical limitations caused by DCD often lead to emotional and behavioral problems such as chronic frustration, low self-esteem, poor motivation, and avoidance of contact with others. After childhood onset, DCD usually persists, causing problems with many of the basic physical skills needed for daily living and work. Children and adults with ASD are more prone to **depression**, which causes them to have frequent episodes of persistent sadness and helplessness. Symptoms include loss of interest in normal activities, sleep disturbances, cognitive impairment, and suicide ideation. People with ASD are also more likely to have **bipolar disorder**, a comorbidity that causes them to cycle between **manic** mood states, during which they are energetic and happy, and depressive mood states, during which they feel sad, empty, and irritable. Diagnosis of bipolar disorder is especially challenging in the context of ASD. Although the symptoms caused by depressive mood states are usually obvious, the symptoms of mania, such as having inflated self-esteem, being unusually talkative, becoming more easily distracted, and engaging in risk-taking behaviors, are often masked by ASD symptoms related to communication, emotional interaction with others, and stereotyped behaviors.

About a third of people with ASD suffer from **epilepsy**, characterized by recurrent seizures that occur because of abnormal electrical activity in the brain. Seizures typically begin during childhood or adolescence, and both their frequency and severity vary among individuals. The occurrence of epilepsy in people with ASD is positively correlated with the severity of their intellectual disability and behavioral problems. The most serious type of seizure is a **grand mal seizure**, during which someone loses consciousness, collapses, and has violent convulsions. After a grand mal seizure, people are often disoriented, tired, and have a severe headache. People with ASD also have **absence seizures** that cause fluttering eyelids, sudden hand movements, leaning forward or backward, and sudden stopping of speech or motion. Loss of consciousness can occur during absence seizures, which brings the risk of injury. The most common seizures experienced by people with ASD are **subclinical seizures**, which do not produce observable symptoms and can only be detected as abnormal brain electrical activity during an EEG.

Many people with ASD have atypical reactions to sensory inputs. Although their sense of sight, sound, touch, smell, and taste is normal, the manner in which their brains process sensory inputs and integrate them with prior experiences is atypical. Sometimes, this sensory effect is severe enough for the diagnosis of a comorbidity called **sensory processing disorder (SPD)**. Children with SPD can be hypersensitive to stimuli, and develop extreme responses to loud noises and a dislike for being touched, or hyposensitive to stimuli, which causes a high pain tolerance and a constant need to touch people or things. Taking in and processing two sensory inputs at the same time, such as looking and listening, is sometimes difficult for those with SPD. Stimuli that seem normal to others can be confusing, irritating, or even painful. SPD also causes problems with language acquisition, the development of motor skills, and socialization.

Most children with ASD have sleep problems, the most common of which are difficulties in settling down to sleep, restless sleep, repeated episodes of waking during the night, prolonged wakefulness, and waking too early in the morning. Although sleep problems can lead to behaviors such as hyperactivity, aggression, anxiety, irritability, and inattentiveness, the occurrence of these ASD behaviors can also cause sleep problems. Other causes of sleep problems include nightmares, sleep apnea, and restless legs syndrome.

Gastrointestinal disorders are common among people with ASD. Some develop **chronic constipation**, a comorbidity associated with bloating, moderate to severe intestinal pain, fever, and headaches, and the potential long-term effects of recurrent vomiting, weight loss, damage to the skin around the anus, and protrusion of the rectum from the anus. Others have **chronic diarrhea**, defined as the persistence of loose or watery stool for at least 4 weeks. Symptoms include abdominal cramps, abdominal pain, bloating, fever, and nausea, which can lead to dehydration, urinary tract infections, kidney failure, and seizures. **Gastroesophageal reflux disorder (GERD)** is also a comorbidity of ASD, and is caused by the acidic mixture of partially digested food from the stomach traveling back into the esophagus, resulting in mild to severe heartburn, chest pain, coughing, regurgitation, and a bad taste in the mouth. A telltale sign of GERD pain is that these immediate symptoms worsen after lying down. Long-term effects of GERD include erosion of teeth, disturbance of sleep, and dietary problems.

Although occasional eating of nonfood items is normal for children under the age of 2 years, some children with ASD continue this behavior, and are diagnosed with an eating disorder called **pica**. Children with pica have an unusual appetite for things that do not have any nutritional value, such as soil, rocks, ice, glass, hair, or feces. They are at risk for ingesting life-threatening poisons or causing severe physical damage to their digestive systems that requires emergency surgery. Long-term complications of pica depend on what is ingested, and include lead poisoning from paint ingestion, pathogenic infections of the digestive system from soil or feces, and intestinal obstruction from indigestible materials.

CHAPTER 2

Causes and Contributing Factors

The basis for all human traits and conditions can be understood in terms of genetics, environment, or a combination of both. Thousands of diseases are known to be caused solely by genetics. Patterns of disease inheritance arise from simple rules for the transmission of **genes** from one generation to the next. Examples include cystic fibrosis, which is caused by inheritance of two faulty copies of the *CFTR* gene, sickle cell disease, which results from the inheritance of two faulty copies of the *HBB* gene, and Huntington's disease, which is caused by **mutation** of the *HTT* gene. Many diseases have no genetic cause, and are wholly caused by environmental influences. Examples include infectious diseases such as malaria, tuberculosis, and hepatitis, and environmental diseases such as skin cancer or chronic obstructive pulmonary disease (COPD). For most diseases, the cause cannot be assigned exclusively to either genetics or environment, and must be understood as an interplay between them. The genetic contribution to these diseases occurs by the inheritance of disease susceptibility. Sometimes, susceptibility occurs primarily because of variation in single genes, as is the case with the *BRCA1* gene. Inheritance of a variant *BRCA1* gene can dramatically increase the risk of breast or ovarian cancer. More commonly, however, disease susceptibility is caused by variations in multiple genes. Diseases that are caused by the interaction of multiple genes with environmental factors are called *complex trait diseases*. Although they cluster in families, complex trait diseases do not follow simple patterns of inheritance. Examples include type I diabetes, heart disease, Parkinson's disease, Alzheimer's disease, and obesity. There are rare cases in which ASD is caused by a single gene, but ASD is primarily a complex trait disease. This chapter describes evidence that susceptibility to ASD is inherited, methods by which ASD susceptibility genes have been discovered, and mechanisms by which mutations affect the function of ASD susceptibility genes.

How Do We Know that ASD Is Heritable?

For many diseases and condition, the basis of heredity is straightforward because they are caused by a single gene and result in a simple inheritance pattern that is easily detected within families. Whether or not an individual has one of these **monogenic** diseases is solely determined by which versions of the disease-causing gene they inherited from their parents. Monogenic diseases are also called Mendelian diseases, a reference to Gregor Mendel, who first published the concept of a gene as a unit of hereditary information in 1866. Mendel discovered that genes can exist in different forms that we call **alleles,** and that an allele can be **dominant** over a **recessive** allele when both are present in the same individual. Descriptions of the alleles that someone inherits are referred to as **genotypes,** whereas the traits, conditions, or diseases encoded by genotypes are called **phenotypes.** Alleles that cause monogenic diseases can be dominant, as is the case for Huntington's disease, which means that inheritance of a single disease-causing allele results in disease, or recessive, as in cystic fibrosis, which means that disease results from the inheritance of two disease-causing alleles. Establishment of the pattern of inheritance for Mendelian diseases enables reliable predictions to be made about the probability of children inheriting a given disease. For complex trait diseases such as ASD, predictions are extremely difficult because these diseases involve multiple genes that make unequal contributions to the phenotype, and are affected by a variety of environmental factors.

The complex interplay between multiple susceptibility genes and various environmental influences makes it difficult to measure the relative contributions of genetics and environment to the risk of ASD. An important experimental approach that addresses this challenge is **twin studies,** which use comparisons between identical twins and fraternal twins to distinguish the effects of genetics and environment. During human reproduction, fertilization occurs when a sperm cell carrying genes from the father joins an egg cell carrying genes from the mother. The **human genome** is carried on 24 different types of **chromosomes**. The X and the Y chromosomes are referred to as the sex chromosomes because they determine human sex. Typical females have two X chromosomes, whereas typical males have one X and one Y chromosome. The other chromosomes are called **autosomes**, and are numbered 1 through 22 in approximate descending order of size. Chromosomes can be easily identified under the microscope by their sizes and the positions of their **centromeres**, which separate them into long and short arms. During fertilization, a sperm cell carrying 23 chromosomes, one of which is either the X or the Y chromosome, fuses with an egg cell carrying 23 chromosomes, one of which is an X chromosome. The resulting fertilized egg cell is called a **zygote**, and it has a full complement of 46 chromosomes. During the first embryonic cell division, the combined genetic material of a newly formed zygote is copied and partitioned to each of two newly formed cells, and this process is repeated over and over to provide all the cells of the developing embryo with equivalent genetic material, aside from the effects of rare spontaneous mutations. Identical twins are produced when the first two embryonic cells separate and develop into two distinct but genetically equivalent embryos. Because these twins originated from a single zygote, identical twins are also called **monozygotic twins**. Identical twins inherit the same alleles from their parents because they were produced by a single fertilization event. Identical twins are always the same sex. Fraternal twins are called **dizygotic twins** because they occur when two fertilization events independently produce two genetically distinct zygotes that develop into two genetically distinct children. The genetic relationship between fraternal twins is the same as between any two siblings, which means they share, on average, half of their alleles. Fraternal twins are just as likely to be of different sexes as they are to be of the same sex. Twin studies exploit the genetic relatedness of monozygotic compared with dizygotic twins, and the shared environmental influences that both types of twins experience. By comparing these two types of twins, scientists measure **heritability**, the degree to which phenotypic variation is due to genetics, as well as **concordance rate**, the probability that a second twin has a given phenotype if the first twin has it. If a trait, condition, or disease is exclusively caused by inheritance, with no influence from environment, heritability is 100 percent. The concordance rate for monozygotic twins is 100 percent because they have the same genotype. The concordance rate for dizygotic twins is determined by whether the phenotype-causing gene is located on an autosome or a sex chromosome, and whether the phenotype-causing allele is dominant or recessive. An example of a trait with 100 percent heritability is ABO blood type. Identical twins always have the same blood type because they share the same chromosomes. The concordance rate for fraternal twins depends on the genotypes of the parents, and can vary from 25 to 100 percent. Many traits, conditions, and diseases are completely determined by environmental factors, which means that heritability is zero, and the concordance rate of phenotypes between identical twins is no higher than it is for fraternal twins. For example, because fingerprints are not genetically determined, everyone has a unique fingerprint, so both the heritability and the concordance rate for fingerprints are zero. Most phenotypes have a heritability between 0 and 100 percent, which means that they are determined by a combination of genetic and environmental causes. For example, twin studies have measured the heritability of IQ to be 45 percent for children and 75 percent for adults.

Twin studies conducted over the past 40 years have yielded estimates of the heritability of ASD that range from 60 to 90 percent, indicating that ASD is largely determined by genetics, but that environmental influences also play a role. Monozygotic twin concordance rates are 70 to 90 percent, whereas dizygotic twin concordance rates are less than 10 percent. The variation in heritability and concordance rate measurements determined by various twin studies comes from differences in sample size, the use of in-person or proxy diagnoses of participants, and differences in the diagnostic tools used. Recent changes in the diagnostic criteria for ASD have also impacted the results of twin studies. Heritability estimates and dizygotic concordance rates are lower for studies that include Asperger's syndrome and PDD as part of ASD, which indicates that the contributions of genetics and environment are probably not constant along the ASD continuum. An important conclusion from ASD twin studies is that nontwin siblings of individuals with ASD are 20 times more likely to have ASD than members of the general population.

Discovery of Genes that Cause ASD

Although they provide evidence of heritability, twin studies are insufficient to identify specific ASD susceptibility genes. One approach to the discovery of ASD susceptibility genes is to conduct a **genome-wide association study (GWAS)**, which compares genome variations in a large population against a common reference genome sequence. All humans share about 99.7 percent of the 3.2 billion DNA bases in the human genome, which means that there are about 10 million places where the DNA base present in one person will be different when compared with the genome of another person. These places are called **single nucleotide polymorphisms**, or SNPs (pronounced "snips"). The term *polymorphism* means many shapes, although in this context there can be only four shapes because there are only four DNA bases. For example, a *C* might be present at a particular position on chromosome 5 for some people, whereas others have a *G* in that position. To qualify as an SNP, an alternative base must be found in at least 1 percent of people.

Over the past decade, multiple research groups have employed GWASs to identify candidate ASD susceptibility genes. For example, a 2009 GWAS involved 438 children and adolescents who had been diagnosed with autism, according to *DSM-IV* diagnostic criteria. The study determined the genotype for over one million SNPs among the participants with autism and members of their immediate families. The investigators uncovered a correlation between variations in 96 SNPs located on various chromosomes and the occurrence of autism, which indicated there are many susceptibility genes for autism. A particularly strong correlation was found between autism and variations in 8 SNPs located on the short arm of chromosome 5. This same chromosomal region was implicated in a 2010 GWAS that examined 500,000 SNPs in 1,031 families that included a child with autism. The study also found SNPs on the long arm of chromosome 6 and the short arm of chromosome 20 that are correlated with autism. The results of these and other GWASs for ASD conducted over the past decade were extended in 2017 by a large-scale GWAS organized as an international collaboration by the ASD Working Group of the Psychiatric Genomics Consortium. The GWAS examined 7 million SNPs among 16,539 children with ASD, as diagnosed by the

ADI-R or the ADOS-G, and 157,234 typically developing children. In addition to previous correlations, new correlations were found between the occurrence of ASD and SNPs on the long arm of chromosome 10, the short arm of chromosome 3, and the short arm of chromosome 8. The net result of all GWAS research is clear evidence that ASD is a complex trait disease caused by variations in multiple susceptibility genes.

Gene Mutations Cause ASD

The human genome carries about 22,000 protein-encoding genes, and the median size of a human gene is about 24,000 base pairs. The process by which genes are expressed to produce proteins is illustrated in Figure 2.1, using the ASD susceptibility gene *NLGN3* as an example. Genes carry information in the form of DNA sequences of four bases, commonly referred to as *G, C, A,* and *T*. Ninety-eight percent of human genes are split into **exons**, which encode the amino acids of proteins, and intervening sequences called **introns**. The average number of exons in human genes is about 10, and their average size is 288 base pairs. The first step in **gene expression** is **transcription**, during which the DNA sequence of a gene is copied into an RNA sequence (see Figure 2.1). **RNA splicing** removes the introns and connects successive exons to make a contiguous protein-encoding sequence called a **messenger RNA (mRNA)**. **Translation** is the process by which an mRNA base sequence is used to direct the synthesis of a protein. There are 20 primary amino acids, and their sequence in a protein determines its structure and function, or dysfunction. The bases of mRNA are translated in groups of three, called **codons**. Because there are four possible bases in each of the three positions of a codon, there are 64 ($4 \times 4 \times 4$) different codons possible. Genetics, biochemistry, and molecular biology experiments in the 1960s revealed which amino acid is encoded by each of the 64 codons. This information is referred to as the **genetic code**. One of the codons is a start codon for the initiation of translation, and three are stop codons that signal the end of translation.

The fundamental mechanism by which susceptibility genes cause a disease or condition is gene mutation. Mutations are caused by **mutagens** that damage DNA, or by errors during **DNA replication**. Mutations can

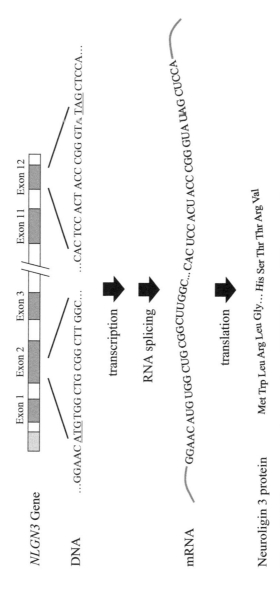

Figure 2.1 Structure and expression of NLGN3, an ASD susceptibility gene

Source: Drawn by Todd T. Eckdahl 2017.

be inherited, or they can arise as **spontaneous mutations** during sperm or egg cells production or embryonic development. The phenotypic effects of mutations can be deleterious, beneficial, or neutral, depending on the function of the gene, the type of mutation, and relationship between the mutant allele and the normal allele. Spontaneous mutation of ASD susceptibility genes is the most common cause of ASD. Researchers estimate that it might be responsible for as many as 95 percent of ASD cases. Mutations have been discovered in many genes that encode proteins involved in the development and function of the brain or central nervous system. The fundamental process by which these systems operate is the transmission of nerve signals between specialized cells called **neurons**. Neurons use ion pumps and ion channels to generate and propagate electrical gradients across their outer membranes that change rapidly to produce an **action potential** when neurons are excited. The connection between two neurons is a **synapse**, which is formed when a long extension of one neuron called an **axon** meets a **dendrite**, a branched extension of a second neuron (see Figure 2.2). When a **presynaptic** neuron is excited, the action potential travels down its axon to the synapse, where chemicals called **neurotransmitters** are released. As illustrated in Figure 2.2,

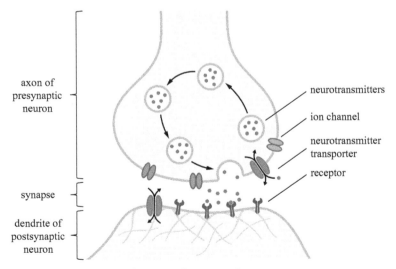

Figure 2.2 Structure of a typical synapse between neurons

neurotransmitters diffuse to fill the synapse and bind to a protein receptor in the membrane of a **postsynaptic** neuron. The effect of neurotransmitter binding on the postsynaptic neuron can be either excitatory or inhibitory. Although there are more than a hundred different neurotransmitters, neurons of the brain frequently use **glutamate**, an amino acid that also functions as an excitatory neurotransmitter. The ability of each neuron in the brain to make synaptic contacts with thousands of other neurons produces an incredibly complex network.

Information about many genes associated with ASD is available in an open access database called **AutDB**. AutDB provides information on 970 genes associated with ASD, including references to the scientific literature. Supporting evidence on the risk associated with these candidate ASD susceptibility genes varies widely, but most researchers agree with the conclusion that there are about 70 ASD susceptibility genes (Table 2.1). *NLGN3* is a susceptibility gene located on the long arm of the X chromosome and contains 12 exons. The gene encodes neuroligin 3, a protein that is produced in large amounts in the brain and central

Table 2.1 Examples of ASD susceptibility genes

Gene	Protein	Protein function
NLGN3, NLGN4X	Neuroligin-3, Neuroligin-4	Interacts with neurexins to control the initial formation of synapses in the brain and central nervous system
NRXN1	Neurexin-1-alpha	Forms complexes with neuroligins at synapses in the brain and central nervous system
FOXP1	Forkhead box protein P1	Transcription factor that controls transcription of genes required for brain, heart, and lung development
SHANK3	SH3 and multiple ankyrin repeat domains 3	Connects neurotransmitter receptor and ion channel proteins on the surfaces of postsynaptic neurons to structural and signaling proteins on the inside of the cells
CNTNAP2	Contactin-associated protein-like 2	Clusters ion channels required for nerve signal transmission along the axons of neurons

nervous system, and in smaller amounts in adrenal glands, ovaries, testes, and the uterus. In the brain, neuroligin 3 is located on the surfaces of neurons, where it contributes to the formation of synapses, such as the one shown in Figure 2.2. One example of a mutation in *NLGN3* is called p.Arg451Cys. The mutation name starts with a *p* for protein, and indicates that amino acid arginine (Arg) at position 451 in the typical protein is replaced with amino acid cysteine (Cys) in the protein associated with ASD. The change in amino acid disrupts the ability of the neuroligin-3 protein to interact with proteins called neurexins. Neurexins are located inside postsynaptic neurons and are required for the proper movement of proteins to the cell surfaces of neurons. The p.Arg451Cys mutation is called a **gain-of-function mutation** because it produces a new function in neurons even when the normal neuroligin 3 protein is present, similar to a dominant allele. Further evidence of the causal relationship between p.Arg451Cys and ASD comes from the observation that mice with p.Arg451Cys display impaired social interactions and aberrant nerve cell communication. *NLGN4X* is another susceptibility gene for ASD, and it spans about 388,000 base pairs on the short arm of the X chromosome, making it a relatively large gene. *NLGN4X* has 21 exons, and it encodes a protein called neuroligin-4 at higher levels in the brain, thymus, and ovary, and at lower levels in a variety of other tissues. One mutation of *NLGN4X* that causes ASD susceptibility is an insertion of a single base into an exon which results in a truncated version of the neuroligin-4 protein that is only 395 amino acids long instead of the normal 816 amino acids. Because the truncated protein cannot properly interact with neurexins to initiate the formation of synapses, this mutation is a **loss-of-function mutation**. Truncated neuroligin-4 protein in mice causes deficits in reciprocal social interaction and communication that mirror the effects of ASD on people.

Susceptibility to ASD is associated with a mutation of the *NRXN1* gene, too. *NRXN1* is located on the short arm of chromosome 2, encodes the neurexin-1-alpha protein, and is one of the largest genes in the human genome with 24 exons that span over 1 million base pairs. Neurexin-1-alpha is a member of the neurexin family of proteins, which interact with neuroligins to form complexes that contribute to the function of neuron synapses (see Figure 2.2). Expression of *NRXN1* occurs primarily in the brain and central nervous system, and results in the production of over 3,000 normal variants of neurexin-1-alpha due to the many different combinations of exons connected during RNA splicing. One loss-of-function mutation in *NRXN1* that confers ASD susceptibility is p.Ser14Leu, which replaces the 14th amino acid, serine, with leucine. Mice lacking *NRXN1* have reduced excitatory nerve transmission and behaviors analogous to ASD in humans.

Another susceptibility gene is *CNTNAP2*, with 33 exons spanning over 2.3 million base pairs on the long arm of chromosome 7. *CNTNAP2* encodes contactin-associated protein-like 2, a form of neurexin that clusters ion channels required for nerve signal transmission along the axons of neurons. In addition to the brain and central nervous system, *CNTNAP2* is expressed in a variety of tissues, including kidney, lung, thyroid, ovary, and testis. One *CNTNAP2* mutation that is associated with ASD is the loss-of-function mutation g.681154A>T, so named because an *A* at position 681,154 in the gene is replaced with a *T*. The mutation occurs in intron 2, and it affects the frequency at which the intron is properly removed by splicing to produce an mRNA that can be translated into contactin-associated protein-like 2. Mice in which the gene has been deleted have absence seizures, and learning deficits similar to human ASD symptoms.

The 600,000 bp *FOXP1* gene on the short arm of chromosome 3 encodes forkhead box protein P1, a transcription factor that controls the expression of genes in cells that become embryonic brain, heart, and lung tissues. The *FOXP1* mutation p.Arg525Ter confers ASD susceptibility and is a loss-of-function mutation. This mutation changes a codon specifying arginine to a stop codon, producing a truncated protein of only 525 amino acids in length instead of the normal 677 amino acids. The missing part of the mutant protein is the part that targets it to the nucleus and enables it to bind to DNA to control the transcription of genes. Mice lacking *FOXP1* have abnormal brain development and display aberrant repetitive behaviors, impaired short-term memory, hyperactivity, and deficits in social behavior.

SHANK3 is also a susceptibility gene for ASD. The 58,000 base pair gene is found on the long arm of chromosome 22, and includes 27 exons. The encoded protein is called SH3 and multiple ankyrin repeat domains

3, and it connects postsynaptic cell surface neurotransmitter receptors and ion channels to structural and signaling proteins on the inside of the cells (see Figure 2.2). *SHANK3* mRNA is produced in the brain and the spinal cord, and in a variety of other tissues, including lung, pancreas, muscle, ovary, and testis. The loss-of-function mutation p.Ala1227Glyfs inserts a *G* nucleotide in codon 1227 and produces a truncated protein of only 1227 amino acids instead of the normal 1731 amino acids. In mice, p.Ala1227Glyfs causes synaptic transmission defects and impaired social interaction, and in people, it is associated with moderate to severe intellectual disability.

Copy Number Variations Cause ASD

Symptoms of ASD arise when loss-of-function or gain-of-function mutations in individual genes disrupt the development and maintenance of neuronal connections in the brain and central nervous system. However, many mutations that cause ASD are **deletions** or **insertions** of as many as millions of DNA base pairs that affect single genes or many genes arrayed in a chromosomal region. Because they alter the number of gene copies on a chromosome, these mutations are called **copy number variations (CNVs)**. Between 5 and 10 percent of the human genome includes CNVs, and they produce more variation in people worldwide than SNPs do.

CNVs that cause ASD can be inherited, but usually they occur spontaneously during the production of sex cells or during early embryonic development. The rate at which new CNVs occur is 3 to 5 times higher among people with ASD than it is in the general population. The discovery of CNVs has been aided by large-scale DNA sequencing analyses of a cohort called the **Simons Simplex Collection (SSC)**, which consists of over 2,500 families with at least one member who has been diagnosed with ASD. SSC research has shown that small CNVs usually contain a single gene associated with a high risk of ASD, whereas larger CNVs usually contain several genes that carry a lower risk of ASD. CNVs were found to span an average of 0.94 genes in individuals with ASD, but only an average of 0.12 genes in their siblings. ASD-associated CNVs are categorized as either recurrent, which means that the same CNVs arise independently in many people with ASD, or nonrecurrent, which means that each CNV is unique. Both recurrent and nonrecurrent ASD-associated CNVs have been found on each of the 24 chromosomes. The AutDB currently catalogs over 2,200 ASD-associated CNVs. Recurrent CNVs are found at hotspots of genomic instability, where there is a fourfold higher frequency of large deletions and duplications than the rest of the genome. Hotspots contain long stretches of repetitive DNA sequence that contribute to the duplication or deletion of DNA as a result of errors in DNA replication, or in the process by which DNA is exchanged from one chromosome to another during **genetic recombination**. Hotspots establish risk loci for ASD-associated CNVs. Figure 2.3 shows a typical male **karyotype** with 22 pairs of autosomes, an *X* chromosome, and a *Y* chromosome. The arrows indicate the locations of 6 ASD CNV risk loci found on chromosomes 1, 3, 7, 15, 16, and 22 that were identified using the SSC cohort. The names of the locations begin with the number of the chromosome, followed by a *p* for the short arm (from the French, "petite") or a *q* for the long arm. Each ASD CNV risk locus includes at least 13 genes, and among these are independently established ASD susceptibility genes. Gain-of-function mutations that raise the risk of ASD arise by an increase in the copy numbers of these genes, whereas loss-of-function ASD mutations occur by copy number reductions.

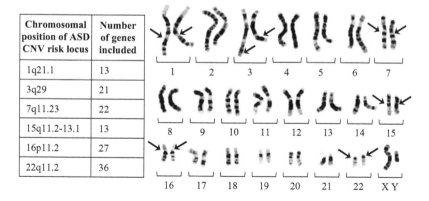

Chromosomal position of ASD CNV risk locus	Number of genes included
1q21.1	13
3q29	21
7q11.23	22
15q11.2-13.1	13
16p11.2	27
22q11.2	36

Figure 2.3 ASD CNV risk loci with number of genes included and positions indicated by arrows on a karyotype

Source: Karyotype image adapted in 2017 by Todd T. Eckdahl from: National Human Genome Research Institute, http://www.genome.gov/glossary/resources/karyotype.pdf.

The most common CNVs in people with ASD, found in 1 to 3 percent of all ASD cases, are recurrent, spontaneous duplications of most or
all the ASD CNV risk locus on the long arm of chromosome 15 (15q11.2
to 15q13.1). The underlying cause of the ASD symptoms associated with
these CNVs is a change in the pattern of expression of genes found in
the region. One of the genes is *UBE3A*, which contributes to the function of synapses by regulating the destruction rate of other proteins, and
several other genes that produce protein receptors for an inhibitory neurotransmitter. Altered patterns of expression of these genes are caused by
disruption of **genomic imprinting**, a process by which the chromosomal
packaging of genes is affected by whether they are inherited from the
mother or the father. The effects of genomic imprinting are illustrated
by deletions of the 15q11.2 to 15q13.1 region. A paternally inherited
deletion of this region results in **Prader–Willi syndrome**, but a maternally inherited deletion causes **Angelman syndrome**, and symptoms
for the two syndromes are distinct. Genomic imprinting is an example
of **epigenetics**, which seeks to understand how the expression of genes
can be controlled by environmental influences that cause changes in the
packaging of genes in chromosomes.

Inheritance of Monogenic Syndromic ASD

Most cases of ASD are caused by spontaneous small mutations or CNVs
that affect multiple ASD susceptibility genes. However, between 2 and 5
percent of ASD cases occur by inheritance of mutations in single genes.
When monogenic inheritance of ASD occurs, it is often in the context
of a genetic syndrome that causes a variety of symptoms, some of which
are similar to the symptoms of ASD, but many of which are not. This
type of ASD is called **monogenic syndromic ASD**. Over 100 monogenic
syndromes cause at least some of the major symptoms of ASD, and
as many as 20 cause symptoms that are in accord with a diagnosis of
ASD (Table 2.2).

Smith–Lemli–Opitz syndrome (SLOS) is a monogenic syndrome
that occurs at a rate of 1 in 20,000 to 60,000 births. SLOS causes a wide
and variable range of symptoms, including microcephaly, low-set-ears,
cleft lip or palate, and extra fingers or toes, as well as sleep disorders,

Table 2.2 Examples of monogenic syndromes that cause ASD

Syndrome	Inheritance pattern	Gene	Protein and function
Smith–Lemli–Opitz syndrome	Autosomal recessive	DHCR7	7-dehydrocholesterol reductase, an enzyme required for cholesterol synthesis
Tuberous sclerosis	Autosomal dominant	TSC1, TSC2	TSC1 (hamartin), a cell adhesion protein; TSC2 (tuberin), a tumor suppressor
Neurofibromatosis type I	Autosomal dominant	NF1	Neurofibromin, a tumor suppressor that affects nerve signal transmission
Fragile X syndrome	X-linked dominant	FMR1	FMRP, a modulator of the expression of other nerve cell genes

atypical reactions to sensory stimuli, and intellectual disability. More than half of SLOS patients meet the criteria for ASD. The cause of SLOS is loss-of-function mutation of the DHCR7 gene on the long arm of chromosome 11, which encodes an enzyme called 7-dehydrocholesterol reductase that synthesizes cholesterol in various tissues. Cholesterol is a normal and necessary **lipid** that modulates the fluidity of cell membranes. Cells in the brain are particularly sensitive to disruption of cholesterol synthesis, because dietary cholesterol cannot pass the blood–brain barrier. Over 100 mutations of DHCR7 have been described, most of which are recessive, loss-of function mutations. This means that SLOS is inherited in an **autosomal recessive** manner. Someone with the disease must have inherited two copies of the recessive allele, and their genotype is called **homozygous recessive**. An individual who has inherited two dominant alleles is **homozygous dominant**, and is unaffected by the disease, whereas someone who has inherited one normal dominant allele and one mutant recessive is **heterozygous**, and can be referred to as a **carrier** if that person does not have any symptoms of the disease. Figure 2.4 illustrates the most common scenario by which autosomal recessive genetic diseases are inherited, which is a mating between two carrier parents. The figure indicates genotypes with a D for the normal dominant DHCR7 allele and a d for the mutant recessive allele. A **Punnett square** helps illustrate the possible

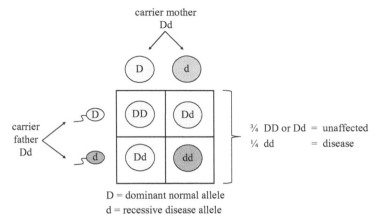

Figure 2.4 *Autosomal recessive inheritance illustrated with a Punnett square showing offspring genotypes and phenotypes from a carrier father and a carrier mother*

Source: Drawn by Todd T. Eckdahl 2017.

offspring genotypes and their frequencies of occurrence. In this example, there is an equal probability that a sperm involved in zygote formation contains the dominant allele or the recessive allele. The same is true for egg cells, so each of the four possible fertilization events is equally likely. Illustrated by the fact that one of the four boxes in the Punnett square includes the homozygous recessive genotype (dd), the probability of a given child inheriting the autosomal recessive disease is ¼, or 25 percent. This probability applies independently to each fertilization event and must be calculated separately for all subsequent pregnancies. For SLOS, this simple scenario is complicated by the occurrence of many different recessive alleles in the population. When an individual inherits two different recessive alleles, that person's genotype is called **compound heterozygous**. The large range of symptom severity among patients with SLOS is due to complex interactions between recessive alleles found in people with compound heterozygous genotypes.

Tuberous sclerosis occurs at a rate of about 1 in 6,000 children, and it causes benign tumors to form in organs throughout the body, such as the brain, kidneys, liver, lungs, skin, and eyes. The symptoms of tuberous sclerosis vary widely, and depend on which organs are affected. Because childhood brain tumors cause developmental delays and behavioral problems such as attention deficit/hyperactivity disorder (ADHD), **obsessive compulsive disorder (OCD)**, aggression, destructive behaviors, and deficits in social interactions, about half of children with tuberous sclerosis are diagnosed with ASD. The cause of tuberous sclerosis is mutation of either of two very similar genes called *TSC1* and *TSC2*. *TSC1* is on the long arm of chromosome 9 and is expressed in the brain and other tissues to produce several variants of a protein known as either TSC1 or hamartin, which plays a role in the control of cell-to-cell adhesion. *TSC2* is found on the short arm of chromosome 16 and is expressed with alternative splicing in the brain and various other tissues. The protein variants function as a tumor suppressor called TSC2 or tuberin that slows the enzymatic cascade leading to cell growth, a prerequisite for cell division. Mutations in either *TSC1* or *TSC2* increase the risk of tumor development, and they occur spontaneously in about one-third of cases, but are inherited in the majority of cases in an **autosomal dominant** manner. Figure 2.5 shows the most common way that tuberous sclerosis is inherited, involving one parent who has the disease and one who is unaffected by it. The Punnett square illustrates the possible offspring genotypes

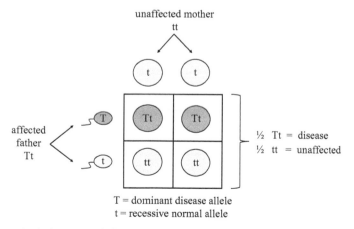

Figure 2.5 Autosomal dominant inheritance illustrated with a Punnett square showing offspring genotypes and phenotypes from a heterozygous affected father and an unaffected mother

Source: Drawn by Todd T. Eckdahl 2017.

produced by a mating between a father affected by the disease who is heterozygous (Tt) and an unaffected mother who is homozygous recessive (tt). In this mating, there is an equal probability that a sperm cell will contain the dominant allele or the recessive allele, whereas egg cells carry only the healthy recessive allele. The probability of a given offspring having the unaffected genotype of tt is ½, or 50 percent. The chance of the offspring having the heterozygous genotype (Tt), and therefore having the disease, is also 50 percent. The reciprocal mating between an affected mother who is heterozygous and an unaffected father gives the same probabilities. A mating between two heterozygous parents, both with the same autosomal dominant disease, would result in a 75 percent chance that an offspring will inherit the disease.

Monogenic syndromic ASD can also be caused by **neurofibromatosis type 1 (NF1)**, a syndrome with a global occurrence of about 1 in 3,500 that causes the growth of benign tumors on or just beneath the skin and along nerve fibers throughout the body. Disruption of nerve function by the tumors can cause seizures, blindness, hypertension, curvature of the spine, and abnormal bone growth. Many people with NF1 display ASD symptoms. Some studies have found that more than one-quarter of patients with NF1 also have ASD. NF1 is caused by mutations in the *NF1* gene on the long arm of chromosome 17. The 88 exons of *NF1* are used to produce many RNA splice variants of a protein called neurofibromin in a myriad of cells. Neurons, and the cells that support them, are affected by mutant neurofibromin because of its critical role in nerve signal transmission, and as a **tumor suppressor**. Neurofibromin normally inhibits the function of a family of protein enzymes called Ras, which is the first protein in a cascade of enzymes leading to the expression of genes that cause the growth and specialization of cells. Ras proteins are called G proteins or GTPases because their activation is controlled by the binding and hydrolysis of GTP. More than 1000 *NF1* loss-of-function mutations have been linked to NF1 disease, including single base changes, small insertions and deletions, and CNVs. Loss of neurofibromin function can occur in cells throughout the body, and when it occurs in specialized cells that support nerve cells, tumors are formed that lead to NF1 symptoms. In about half of NF1 cases, a mutation in *NF1* arises spontaneously in a family, but in the other half, a mutant *NF1* allele is inherited from a parent. People born with one nonfunctional *NF1* allele almost always acquire spontaneous mutations of their one remaining functional allele, which results in a delayed onset of NF1 disease. NF1 is considered to have an autosomal dominant pattern of inheritance because half of the children will develop NF1 if they have one parent with the disease. However, because both *NF1* alleles must become nonfunctional to cause disease, the molecular genetic disease mechanism of NF1 is more akin to autosomal recessive diseases.

Fragile X syndrome (FXS) occurs at a rate of about 1 in 4,000 males and 1 in 8,000 females, and causes characteristic physical features such as a long and narrow face, large ears, a prominent jaw line, and atypically flexible fingers. Children with FXS have behavioral problems such as anxiety, hyperactivity, and attention deficit disorder (ADD). They also show varying degrees of intellectual disability, with boys more affected than girls. About one-third of children with FXS also exhibit symptoms of ASD. The cause of FXS is mutation of the *FMR1* gene on the long arm of the *X* chromosome. Expression of *FMR1* occurs at a high level in the brain and at lower levels in throughout the body. Alternative splicing of the mRNA produces at least 20 variants of a protein called FMRP, which modulates the translation of a variety of mRNAs in neurons that are needed for synapse formation. When *FMR1* is mutated, uncontrolled production of proteins leads to neuronal dysfunction and causes the symptoms of FXS. The normal version of the *FMR1* gene contains a feature called a **trinucleotide repeat**, which consists of successive occurrences of CGG. The number of copies of CGG in the trinucleotide repeat varies widely among people, and ranges from 5 to over 1,000. The mutation of *FMR1* that is responsible for FXS is called a **trinucleotide repeat expansion** because the number of CGG repeats increases over successive generations. The mechanism by which this increase affects gene expression is epigenetic because the *C* in the sequence is a target for **DNA methylation**, which results in tighter packaging of chromosomal DNA and the repression of gene expression. An increase in the number of CGG repeats increases the number of methylated C bases, which reduces *FMR1* transcription. The increased methylation causes a

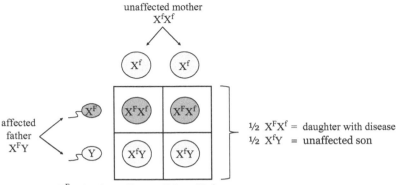

X^F = dominant disease allele on X chromosome

X^f = recessive normal allele on X chromosome

Y = Y chromosome

Figure 2.6 X-linked dominant inheritance illustrated with a Punnett square showing offspring genotypes and phenotypes from an affected father and an unaffected mother

Source: Drawn by Todd T. Eckdahl 2017.

constriction of the X chromosome that makes it appear bent and thus fragile under the microscope, giving FXS its name. The pattern of inheritance of FXS is called **X-linked dominant**. For X-linked diseases, genotypes are expressed using the letter X with a superscript for the allele on that chromosome. The symbol for a Y chromosome is also used in males, but it is not modified with a superscript because it does not contain a copy of the disease-causing allele. Figure 2.6 illustrates a mating between a father who has an X-linked dominant disease such as FXS (X^FY) and an unaffected mother (X^fX^f). The overall probability of offspring inheriting the disease is 50 percent because all daughters will receive the disease-causing X^F chromosome from their father and will therefore inherit the disease, whereas all sons will inherit a Y chromosome from their father and will be unaffected. Figure 2.7 shows that the reciprocal mating of a mother who has an X-linked dominant disease and an unaffected father would turn out differently. The mother is likely to be heterozygous (X^FX^f) because of the rarity of the disease, so the probability that she will pass on an X chromosome that would cause disease in either male or female offspring is 50 percent.

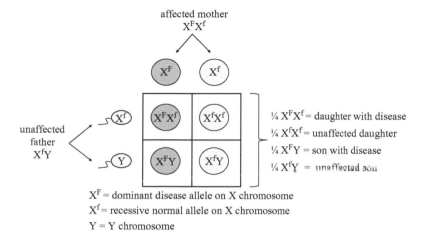

X^F = dominant disease allele on X chromosome
X^f = recessive normal allele on X chromosome
Y = Y chromosome

Figure 2.7 X-linked dominant inheritance illustrated with a Punnett square showing offspring genotypes and phenotypes from an unaffected father and an affected mother

Source: Drawn by Todd T. Eckdahl 2017.

Genetic Testing for ASD

We know the categories of molecular genetic abnormalities that bring about the wide range of types and severity of ASD symptoms to be mutations in single genes that have large effects, mutations in multiple genes that have additive small effects, and chromosomal rearrangements that increase or decrease the copy numbers of genes. We have learned the identities of many ASD susceptibility genes and the nature of the loss-of-function or gain-of-function mutations that cause their dysfunction, and have found many CNVs that are associated with ASD. Our knowledge enables us to develop genetic tests for ASD that contribute to diagnosis of ASD, along with family history, symptoms, and clinical tests. Identifying the genetic cause of ASD also enables prediction of the likely progression of the disease that can be used to develop a personalized plan of treatment. A genetic diagnosis of ASD can be a source of comfort for patients and families, but it can also cause strong emotional responses. Genetic counselors can help patients and their families understand and cope with genetic test results.

The detection of ASD susceptibility gene mutations begins when DNA is extracted from a sample of cells taken from an ASD patient. The DNA sample is subjected to the **polymerase chain reaction (PCR)**, which amplifies specific DNA segments. The resulting PCR products can be analyzed by gel electrophoresis, which measures the size of DNA molecules, to detect insertion or deletion mutations. Alternatively, **DNA microarrays** can be used to simultaneously screen for the presence or absence of thousands of possible mutations in ASD susceptibility genes (see DNA Microarray Methodology URL in Bibliography). PCR products of a patient DNA sample are labeled with a fluorescent tag and allowed to base-pair with DNA fragments carrying known mutations that are arrayed on a solid surface. The fluorescent tag lights up the DNA spot on the microarray that carries the mutation contained in the clinical DNA sample. **DNA sequencing** of PCR products from patient DNA determines the sequence of bases for comparison to known alleles in target genes. If genetic testing for specific ASD susceptibility gene mutations is negative, or if the symptoms and clinical test results are not in accord with the expected effects of known mutations, a form of genetic testing can be performed that detects any mutation in any gene in the entire human genome. **Whole-genome sequencing** is the process by which the sequence of the 3.2 billion base pairs inherited from each parent is determined. Sophisticated computer programs sift through massive amounts of DNA sequence data to find causative mutations for the clinical symptoms. Because most mutations that cause ASD affect the amino acid sequences in proteins encoded by exons, it is often unnecessary to sequence the entire human genome to find them. It is cheaper and faster to do **exome sequencing** because exons comprise only about 1 percent of the total human genome. Large ASD-causative CNVs can be detected by karyotype analysis, or **fluorescent *in situ* hybridization (FISH)**, a method that uses fluorescent DNA probes to detect chromosomal rearrangements. Smaller CNVs associated with ASD are identified by whole-genome sequencing. Genetic testing can determine the underlying cause of many ASD cases, but because our knowledge of ASD is incomplete, 75 percent of ASD cases remain **idiopathic**, which means the underlying cause is still unknown.

What are Contributing Factors for ASD?

A better understanding of the contributing factors for ASD would help explain both idiopathic ASD and the wide variation in symptoms among people with ASD. Many environmental influences might contribute to ASD by epigenetic mechanisms that impact chromosome structure and the level of expression of ASD susceptibility genes, including DNA methylation, **histone** modification, and **noncoding RNA** production. Some epigenetic influences occur during pregnancy, when the brain and the central nervous system develop according to genetic instructions that are influenced by the physiology of the mother. Additional epigenetic changes occur during the production of sex cells in parents. Some studies have shown a correlation between the age of parents and ASD, which might be explained by impairment of genomic imprinting of ASD susceptibility genes in sperm or egg cells. Other research has implicated maternal schizophrenia, depression, and anxiety as contributing factors for ASD, which might exert their influence by epigenetic mechanisms during embryonic development. The use of certain medications during pregnancy also increases the risk of ASD. For example, the epilepsy and bipolar medication, **valproate**, increases the occurrence of ASD. A 50 percent higher risk of ASD is also associated with maternal diabetes. Explanations for this include toxic effects of high glucose levels in the blood on the developing fetal brain, reduction in the availability of oxygen to the fetus, and an increase in oxidative stress. The effects of maternal diabetes are further increased by obesity. Bacterial or viral infections during pregnancy, such as measles, mumps, influenza, pneumonia, and syphilis, are also contributing factors for ASD. These infections are thought to disrupt brain development by stimulating an inflammatory response by the fetal immune system.

The increased public awareness of ASD has been accompanied by harmful misconceptions about the contributing factors for ASD. One of the misconceptions can be traced back to speculative opinions expressed by Leo Kanner in 1943, and echoed in more recent publications, that described the mothers of children with ASD as lacking warmth and the fathers as not being actively engaged with their children. The idea that "refrigerator mothers" or "refrigerator parents" are at least partially responsible for the occurrence of ASD filled a void from the 1950s to the 1980s in the public understanding of ASD that was left by the lack of information on the genetic basis for ASD. There is no evidence, nor has there ever been, that parenting practices cause ASD. Another popular misconception is that the measles, mumps, and rubella (MMR) vaccine increases the occurrence of ASD, an idea that originated with a publication in 1998 by Andrew Wakefield in *The Lancet*, one of the oldest and best known general medical journals in the world. The claim made by Wakefield was fraudulent, and the paper was retracted by *The Lancet* in 2010. Wakefield was found by the General Medical Council to be guilty of professional misconduct and was barred from practicing medicine in the United Kingdom. Despite the fact that no valid scientific link exists between the MMR vaccine and ASD, the urban myth of the manufactured case put forth by Wakefield has led many parents to choose to withhold MMR vaccines for their children, which predictably has led to an increase in infectious diseases, resulting in preventable illnesses and deaths.

CHAPTER 3

Treatment and Therapy

The wide range of symptoms and health complications caused by ASD means that each case must be evaluated individually to determine the most effective course of treatment. There are no medications to cure ASD or treat its core symptoms of impaired social interaction, communication problems, and restricted repetitive behaviors. However, early interventions have been developed for infants and toddlers with ASD that modulate their developmental delays and reduce the severity of symptoms that arise later. Helpful medications are available that enable children and adults with ASD to function better despite the challenges and limitations of their core symptoms, and to treat the wide range of health complications that ASD brings. Effective therapies are available that reduce the core symptoms and health complications of ASD and improve the skills that people with ASD need to function at home, school, or work.

Early Intervention and Education for Children with ASD

The continuous wiring and rewiring of nerve cell connections in the early developing brain of an infant gives it a **plasticity** that gradually transitions to a more stable state later in childhood. Early interventions are ASD therapies that exploit the natural plasticity of the brains of infants and young children. These interventions reduce the effects of developmental delays in the five general categories of physical, cognitive, communication, social or emotional, and adaptive development. Multiple interventions are available, and the variation of ASD symptoms means that each child must be carefully assessed to determine which are most appropriate.

In the United States, families can get access to early intervention services funded by the government. The U.S. Education for All Handicapped Children Act of 1975, renamed the **Individuals with Disabilities Education Act (IDEA)**, ensures that students with a disability are provided with free and appropriate public education that is tailored to their individual needs. IDEA was modified in 1986 to enhance the development of infants and young children with disabilities, reduce the cost of education through early intervention, minimize institutionalization, and help families meet the needs of their children with disabilities. IDEA makes federal assistance available to states for the administration of these early intervention programs.

IDEA also requires free and appropriate public education from age 3 years through high school for all children in an environment that is as inclusive as is practical. There are provisions in IDEA that ensure that children with ASD are not separated into special education classes, or forced to enroll in separate schools, unless the parents and teachers of the child agree that separation is in the child's best interest. IDEA stipulates that children with disabilities, including those with ASD, should be provided services such as transportation, nursing, counseling, and any other support they need to access free and inclusive public education. Because there is considerable variation in the extent to which ASD causes developmental delays and cognitive disability, parents and teachers must establish individualized educational goals for each child, and readjust the goals over time. IDEA mandates the formation of a team composed of parents or other caregivers, teachers, and child development specialists whose responsibility is to develop an **individualized education program** tailored to the educational needs of a given child. Although IDEA requires the team to consider placing the child in the least restrictive environment possible, such as an inclusive classroom with typically abled children, it also allows for the consideration of special needs programs or schools. Efforts to improve the educational opportunities of children with ASD throughout the world are underway, and are supported by ASD organizations such as the Autism Society of America, Autism Speaks, the National Autism Association, and the National Autistic Society (URLs in Bibliography).

Therapies for ASD Core Symptoms and Health Complications

A variety of therapies can treat the core symptoms of ASD and the variety of health complications it brings. One of the most commonly used therapies for ASD is **applied behavior analysis (ABA)**, which uses positive reinforcement to encourage desired behaviors in communication, play, and social interaction. Although ABA is effective as an early intervention, it is also effective for use in older children, adolescents, and adults. ABA can take the form of **discrete trial training**, during which a skill or a concept is taught to a child using a series of simplified scripted exchanges and an established curriculum of interactions. ABA can also take the form of **pivotal response treatment (PRT)**, which is used to improve language and social communication skills. PRT is based on the concept that behaviors associated with communication are pivotal, and affect a wide range of other behaviors. During PRT, a trained professional uses reinforcements and motivations to encourage improvement in communication and social behavior. Another effective practice is a special education program called **training and education of autistic and related communication handicapped children (TEACCH)**, which uses structured teaching in a classroom setting to provide an environment in which children with ASD can learn. TEACCH methods include strict scheduling of activities, the use of clear and concise language, and clearly defined reinforcements. Another commonly used ASD therapy is the **developmental individual-difference relationship-based model**, also known as floortime, during which a therapist engages in play with a child, while emphasizing improvements in emotional and intellectual development. **Cognitive behavior therapy (CBT)** for ASD seeks to change emotions and behaviors by changing the ways that children and adults think. A series of structured sessions clarifies the relationship between thoughts and feelings, and challenges beliefs that underlie undesired behaviors. CBT is effective in reducing anxiety, managing anger, and improving daily living skills in people with ASD. Another therapy is the **picture exchange communication system**, during which a child is encouraged to initiate communication by offering a picture in exchange for something the child wants, such as a toy. The use of picture exchange for communication jump-starts verbal language acquisition.

Therapy for ASD can also take the form of speech and language therapy, which helps children with ASD manage pedantic speech, echolalia, perseverative speech, pronoun reversal, and the use of idiosyncratic and metaphorical language. Occupational therapy helps children develop skills they need to become as independent as possible. Young children can learn personal care skills, such as feeding, dressing, and grooming themselves. School-age children learn skills such as handwriting and keyboarding. The problems that many children with ASD have with sensory input can be addressed with **sensory integration (SI)** therapy. SI therapists use structured play activities to change the way that a child reacts to sights, sounds, textures, and tastes. Some children with ASD require a hearing aid that receives sound with a microphone, converts it into a digital signal, amplifies and adjusts the signal, and transmits sound to the eardrum with speakers. Others require pharmaceutical or surgical treatments for vision problems such as cataracts or corneal dystrophy.

Treatment of ASD Symptoms with Medications

Because no medication can treat the core symptoms of ASD, the use of medications plays a minor role in ASD treatment. However, there are some medications that help manage some of the more severe behavioral problems that children and adults with ASD exhibit. For example, **risperidone** was approved by the U.S. Food and Drug Administration (FDA) in 2006 for the treatment of severe irritability displayed by children with ASD who are at least 5 years old. Risperidone reduces social withdrawal, hyperactivity, stereotyped behaviors, and inappropriate speech. Its adverse effects include dizziness, fatigue, and weight gain. The antipsychotic drug **aripiprazole** is most often used to treat schizophrenia, bipolar disease, and depression, but it was approved by the FDA in 2009 for the treatment of ASD. Aripiprazole reduces irritability, hyperactivity, and the frequency of stereotyped behaviors among children with ASD. Its side effects include insomnia, nausea, increased appetite, and increased susceptibility to respiratory infections. Impairments of social and emotional behavior displayed by children and adults with ASD can be treated with **oxytocin**, a naturally occurring hormone that functions as a neurotransmitter and normally plays a role in sexual reproduction, childbirth, lactation, relationship formation, and social interaction. The most popular method of administering oxytocin is with a nasal spray, and its common side effects include nausea, vomiting, loss of appetite, and memory loss. The psychostimulant, **methylphenidate**, is effective in the treatment of ASD hyperactivity, attention deficits, and impulsivity, but can produce side effects such as nervousness, anxiety, stomach pain, and nausea. Repetitive behaviors and obsessive–compulsive behaviors can be reduced by treatment with antidepressant drugs such as **fluoxetine**, which is sold as Prozac. Fluoxetine is called a **selective serotonin reuptake inhibitor (SSRI)** because it limits the reabsorption of the neurotransmitter serotonin. Common adverse effects of fluoxetine treatment include insomnia, headaches, nausea, and diarrhea. Cognitive impairments in children with ASD can sometimes be improved by treatment with **memantine**, which is called a **glutamatergic** agent because it blocks nerve impulses that use the excitatory neurotransmitter glutamate. Side effects of memantine include anxiety, aggression, increased heart rate, dizziness, nausea, weight loss, and joint pain. Seizures that are caused by ASD can be treated with anticonvulsants such as valproate, whose side effects include nervousness, depression, insomnia, nausea, stomach pain, and fever.

ASD Organizations

Organizations throughout the world promote public awareness of ASD, offer information, resources, or services to people with ASD, and fund ASD research, including Autism Speaks, the National Autism Association, and the National Autistic Society (URLs in Bibliography). The oldest and largest grassroots ASD organization is the Autism Society of America (ASA), founded in 1965 by Drs. Bernard Rimland and Ruth Sullivan, along with other parents of children with autism. With more than 50,000 members and 200 chapters throughout the United States, the ASA increases public awareness of ASD, advocates for ASD programs and services, and provides information and referral services for people with

ASD. Autism Speaks was founded in 2005 by Suzanne and Bob Wright to support people with ASD and their families, to increase understanding and acceptance of people with ASD, and to fund ASD research. Autism Speaks promotes **Light It Up Blue**, an annual event on World Autism Awareness Day (April 2), during which landmark buildings and bridges throughout the world are illuminated with blue lights to bring awareness to ASD.

CHAPTER 4

Future Prospects

ASD research io ongoing in clinics, hospitals, universities, medical institutes, and government laboratories throughout the world to better understand ASD. One outcome of this research is the discovery of better ways to diagnose ASD, which would make the benefits of appropriate early intervention available to more children. Research is also likely to uncover more of the details about how changes in the sequence and expression of susceptibility genes lead to ASD. From better understanding, new pharmaceutical interventions might be discovered that address specific gene dysfunctions in individual ASD cases.

Searching for Better ASD Biomarkers

A **biomarker** is a physical, chemical, or biological measurement taken from outside a patient that accurately and reproducibly identifies a disease state within a patient. Reliable biomarkers enable earlier disease diagnosis and evaluation of treatment efficacy. Biomarkers can be simple, such as body temperature or pulse, or they can be more complicated, such as the level of mRNA produced by the transcription of a specific gene. There is a need for biomarkers for making an accurate diagnosis of ASD, for determining the effectiveness of existing ASD treatments and therapies, and for conducting research on the underlying causes of ASD. One promising ASD biomarker detects errant metabolic chemicals in blood samples from patients with ASD. The validity of this approach is supported by a 2017 study that quantified chemical intermediates in a folate pathway and a transsulfuration pathway. The study correctly distinguished over 97 percent of its 83 participants with ASD from 76 participants who did not have ASD. A less invasive biomarker approach uses electronic brain imaging to look for structural abnormalities caused by ASD. **Magnetic resonance imaging (MRI)** allows detailed measurements of various parts of the brain, and can detect subtle deviations in the brains of young children with ASD. Abnormalities in the connectivity among the parts of the brain that arise in children with ASD can be detected by **diffusion tensor imaging**, which measures fluid flow in the brain. Researchers are also searching for biomarkers that reflect underlying defects in neurotransmitter production. Real-time measurement of neurotransmitter levels in patients with ASD is challenging, but inroads are being made with **in vivo magnetic resonance spectroscopy (MRS)**, a form of MRI. MRS can also measure metabolites needed for nerve fiber production and chemicals associated with energy metabolism that occur at atypical levels in people with ASD.

ASD Experimental Drugs

Scientists use methods from disciplines such as biochemistry, cell biology, developmental biology, physiology, and genetics to develop experimental animal models of ASD to discover or invent promising drug candidates that can be studied with human clinical trials to test their efficacies. For example, a drug called **suramin** used to treat parasitic diseases such as African sleeping sickness and river blindness was shown to reduce behavioral and neuropathological abnormalities in a mouse model of ASD. Suramin is thought to act by reducing the production of metabolites that are needed for the abnormal and disease-causing persistence of the **cell danger response**, which is stimulated in newborn mice by mimicking a viral infection in their pregnant mothers. In placebo-controlled experiments, suramin improved the social behavior of the mice, as measured by the frequency at which they chose to interact with other mice, and their motor coordination while walking along a rotating rod. Because the cell danger response is more active in people with ASD than those without it, suramin was tested in a double-blind, placebo-controlled human clinical trial to assess its safety and efficacy in treating ASD. Although the trial tested only 10 boys, some of the participants showed improvements in behavior, language, and social interactions, and these results have spurred further investigation.

Another approach for ASD pharmacological intervention is based on knowledge of the function of the *KCTD13* gene, which is located in the 16p11.2 hotspot of genomic instability and has been found to be either duplicated or deleted in many ASD patients. *KCTD13* encodes a protein called potassium-channel-tetramerization-domain-containing 13 (KCTD13), which is involved in protein degradation. In neurons, KCTD13 directs the degradation of RhoA, a GTPase that participates in neuronal synapse formation. Mice that carry a deletion of one of their two copies of *KCTD13* develop brain neurons with fewer synaptic connections and a reduction in brain synaptic transmission. Because researchers hypothesized that these abnormalities were caused by high levels of RhoA, they treated slices of mouse brains with two different drugs known to inhibit the function of RhoA. One inhibitor is a small molecule called **rhosin** and the other is a modified bacterial protein called **exoenzyme C3 transferase**. Because the inhibitor treatments quickly restored synaptic transmission in the brain slices, the researchers are pursuing additional experiments using whole mice. If the results continue to be encouraging, the transition to human trials for treating ASD would be made easier by ongoing clinical trials for the treatment of spinal cord injuries.

Personalized Medicine for ASD

The genomics era of biological research continues to have a dramatic impact on the practice of medicine. The international effort to sequence the human genome, culminating with the announcement of the sequence in 2003, improved DNA sequencing and lowered its cost to the point where it is now feasible to collect genomic DNA sequence information from individual patients. Methods to assess gene expression on a genomic scale by measuring mRNA and protein levels have also become feasible for clinical use. Genomic information can lead to a tailored course of treatment not just to a particular disease, but to a particular patient as well. This approach is called **personalized medicine**, and it has already been developed for some cancers. Personalized medicine for ASD would rely on the collection of genomic DNA sequences from thousands of patients with ASD. DNA data must be integrated with data on the efficacy and safety of specific drug treatments for specific genotypes, which could lead to the rationale design of new drugs tailored to specific dysfunctions that cause ASD. For example, if an ASD patient has incorrect levels of the neurotransmitter serotonin, then a personalized medicine approach could result in the use of an SSRI such as fluoxetine. Alternatively, if the cause of ASD involves the neurotransmitter glutamate, then a glutamatergic drug such as memantine might be more appropriate. An important contribution to the realization of personalized medicine for ASD is an open access database of genomic DNA sequences from thousands of ASD patients and their family members that is curated by Autism Speaks, SickKids, and Verily. The name of the database, **MSSNG**, is pronounced "missing," but has deliberately missing vowels to represent missing answers to questions about the underlying genetic causes of ASD. MSSNG already contains more than 7,000 genomes, along with information about the types and severity of ASD symptoms experienced by individual ASD patients.

Conclusion

We have just begun to appreciate the extent to which ASD impacts individuals, families, and societies. With a currently estimated prevalence of about 1 in 68 people, ASD is more common than we recently thought, and the range of its effects on people is wider. We are beginning to understand the genetic, molecular, cellular, and physiologic dysfunctions that cause the wide range of symptoms and health complications in children and adults with ASD. We know some of the major ASD susceptibility genes, and have developed methods to discover others. Mutations have been identified that change patterns of expression of these genes in the brain and central nervous system and lead to ASD. Large and small chromosomal rearrangements have been discovered that alter gene copy number and cause ASD. The growing database of ASD mutations and CNVs steadily improves our diagnosis of ASD, making early interventions more effective and giving more ASD patients access to treatments and therapies. Scientists are cataloging genetic variations in ASD patients, and correlating them with ASD symptoms. Efforts are underway to collect genomic DNA sequence information from thousands of ASD patients throughout the world, and to curate and disseminate data on the efficacy and safety of specific drug treatments for specific genotypes. The goal is to use knowledge of biochemical and cellular mechanisms of drug action to rationally design new ASD treatments. ASD is a good candidate for personalized medicine, an approach that seeks to tailor pharmaceutical interventions to the genotype that an individual carries. These advances point toward a hopeful future in which a child who *prefers to play* alone is quickly and accurately diagnosed with ASD, and benefits from a personalized plan of safe and effective treatments and therapeutic interventions.

Glossary

absence seizures. Mild seizure that cause fluttering eyelids, sudden hand movements, leaning forward or backward, sudden stopping of speech or motion, and sometimes loss of consciousness.

action potential. A temporary shift in the electrical potential across the membrane of a nerve cell that occurs when it is excited.

agenesis of the corpus callosum (ACC). A brain defect in which there is no communication between the left and right hemispheres of the brain.

allele. One of several possible forms of a gene that differ in nucleotide sequence.

Angelman syndrome. A neurological genetic disease caused by a maternally inherited deletion of part of chromosome 15.

anxiety disorder. Causes people to be in a sustained state of fear about the future and to have heightened stress responses to perceived danger.

applied behavior analysis (ABA). A commonly used therapy for ASD that uses positive reinforcement to encourage desired behaviors in the areas of communication, play, and social interaction.

aripiprazole. An antipsychotic drug, most often used to treat schizophrenia, bipolar disease, and depression, which was approved by the Food and Drug Administration (FDA) in 2009 for the treatment of ASD.

Asperger's syndrome. A developmental disorder characterized by severely dysfunctional social interaction and nonverbal communication and restricted and repetitive patterns of behavior and interests.

attention deficit/hyperactivity disorder (ADHD). A neurological disorder characterized by an ongoing pattern of inattention and/or hyperactivity–impulsivity that interferes with functioning or development.

AutDB. A public database of ASD susceptibility genes and ASD-associated copy number variations (CNVs).

autism. A neurodevelopmental disorder characterized by challenges with social skills, repetitive behaviors, speech, and nonverbal communication.

Autism Diagnosis Interview-Revised (ADI-R). A structured interview that measures reciprocal social interaction, communication, and restricted and repetitive interests and behaviors shown by children aged 2 years and older.

Autism Diagnostic Observation Schedule-Generic (ADOS-G). Assesses social interaction, communication, play, and imaginative use of materials for diagnosis of AD or other developmental disorders.

autism spectrum disorder (ASD). A group of neurodevelopmental disorders characterized by challenges with social skills, repetitive behaviors, speech, and nonverbal communication.

autosomal dominant. A pattern of inheritance in which a dominant allele for a gene located on an autosomal chromosome causes the disease or condition.

autosomal recessive. A pattern of inheritance in which one form of a gene is dominant over another when both are present in the same individual.

autosomes. All of the human chromosomes except the X and the Y chromosomes.

axon. A long extension of a nerve cell that carries an action potential to a synaptic connection with another cell.

biomarker. A physical, chemical, or biological measurement taken from outside a patient that enables clinicians and researchers to accurately and reproducibly identify a disease state within a patient.

bipolar disorder. Causes people to cycle between manic mood states, during which they are energetic and happy, and depressive mood states, during which they feel sad, empty, and irritable.

calendrical savants. A type of savant syndrome characterized by the ability to quickly calculate the day of the week for any date in history.

carrier. A description of genotype meaning that someone is heterozygous for an autosomal recessive disease.

centromeres. The connection between two copies of a chromosome that separate them into long and short arms.

Childhood Autism Rating Scale-2 (CARS2). A tool for the diagnostic assessment of relationships, verbal and nonverbal communication, and emotional responsiveness.

chromosomes. Complexes of DNA and protein in the nucleus of cells that carry genetic information.

chronic constipation. Associated with bloating, moderate to severe intestinal pain, fever, headaches, recurrent vomiting, weight loss, damage to the skin around the anus, and protrusion of the rectum from the anus.

chronic diarrhea. Associated with abdominal cramps, abdominal pain, bloating, fever, nausea, urinary tract infections, kidney failure, and seizures.

Clinical Evaluation of Language Fundamentals (CELF-5). Used to measure spoken language expression and understanding, and to clarify the severity of intellectual disability in children with ASD.

cognitive behavior therapy (CBT). A series of structured sessions used to clarify the relationship between thoughts and feelings, and challenge beliefs that underlie undesired behaviors in people with ASD.

comorbidities. Diseases or conditions that are simultaneously present in a patient.

compound heterozygous. A genotype composed of two different recessive alleles.

Comprehensive Assessment of Spoken Language (CASL-2). Used to measure spoken language expression and understanding.

concordance rate. The probability that a second member of a pair of twins has a given phenotype if the first member has it.

copy number variations (CNVs). Duplications or deletions of segments of chromosomal DNA ranging from hundreds to millions of base pairs that affect single genes or many genes arrayed in a chromosomal region.

deletions. Mutations characterized by the loss of small or large segments of DNA.

dendrites. Branched extensions of nerve cells that form synapses with axons of other neurons.

Denver Developmental Screening Test (DDST). Measures the ability of pre-school children to carry out tasks in the four categories of social contact, fine motor skill, gross motor skill, and language.

depression. A disorder that causes frequent episodes of persistent sadness and helplessness.

developmental coordination disorder (DCD). A delay in the development of motor skills and impairment of physical coordination.

Diagnostic and Statistical Manual of Mental Disorders (DSM). A publication by the American Psychiatric Association that is used for the classification of mental disorders.

Developmental Individual-difference Relationship-based model. An ASD therapy, also known as floortime, during which a therapist engages in play with a child while emphasizing improvements in emotional and intellectual development.

diffusion tensor imaging (DTI). A neuroimaging method that measures the flow of water in the brain.

discrete trial training (DTT). A type of applied behavior analysis during which a skill or a concept, such as color or a shape, is taught to a child using a series of simplified scripted exchanges and an established curriculum of interactions.

dizygotic twins. Fraternal twins, produced by two fertilization events.

DNA methylation. A chemical modification of DNA involving the addition of a methyl group to cytosine that is usually associated with the repression of gene expression.

DNA microarrays. Used to simultaneously screen for the presence or absence of thousands of possible mutations.

DNA replication. The process by which DNA is copied.

DNA sequencing. Any of several methods by which the base sequence of DNA is determined.

dominant. An allele that produces a phenotype when present with a recessive allele in a heterozygous individual.

echolalia. The frequent and often meaningless repetition of the words or phrases spoken by others.

epigenetics. The study of heritable effects of chemical modification of DNA histone proteins on gene expression.

epilepsy. A chronic condition characterized by recurrent seizures that occur because of abnormal electrical activity in the brain.

exoenzyme C3 transferase. An inhibitor of RhoA, a GTPase that is required for the formation of synapses.

exome sequencing. DNA sequencing of only the protein-encoding exons in the genome of an individual.

exons. DNA information in genes interspersed with exons that code for amino acids.

fragile X syndrome (FXS). A monogenic disorder characterized by anxiety, hyperactivity, attention deficit disorder, and intellectual disability.

frameshift mutation. An insertion or deletion that disrupts the translational reading frame that establishes how mRNA bases are read in groups of three during translation.

gain-of-function mutation. A mutation that results in a new gene function that can be pathogenic.

gastroesophageal reflux disorder (GERD). An abnormal reflux of acidic stomach contents.

gene expression. The process by which the information of a gene is used to carry out a cellular function.

genes. The fundamental unit of heredity, encoded as the sequence of bases in DNA.

genetic recombination. The process by which DNA is exchanged from one chromosome to another during sexual reproduction.

genome-wide association study (GWAS). A research approach to the discovery of DNA variations that affect a disease or condition.

genomic imprinting. An epigenetic mechanism by which chromosomal packaging of gene affects its expression.

genotype. The forms of genes that an individual carries.

Gilliam Autism Rating Scale-Second Edition (GARS-2). A diagnostic tool that gathers information about characteristic behaviors associated with ASD.

glutamate. The most commonly used excitatory neurotransmitter in the brain.

glutamatergic. Refers to the use of the neurotransmitter, glutamate.

grand mal seizure. A severe seizure that causes loss of consciousness, collapsing, and violent convulsions.

heritability. The degree to which phenotypic variation is due to genetics.

heterozygous. A genotype that includes one dominant and one recessive allele.

histone. One of several basic proteins that interact with DNA to form chromosomes.

homozygous dominant. A genotype composed of two dominant alleles.

homozygous recessive. A genotype composed of two recessive alleles.

human genome. The approximately 3.2 billion DNA bases found on all the nuclear chromosomes and in the mitochondria of a human being.

hypercalculia. The ability of some people with savant syndrome to very quickly add, subtract, multiply, and divide large numbers.

idiopathic. Refers to a disease or condition for which the underlying cause is unknown.

idiosyncratic language. A pattern of speech during which people make reference to words and phrases that do not make sense without knowing where they came from.

in vivo magnetic resonance spectroscopy (MRS). A type of magnetic resonance imaging that enables the measurement of changes in metabolite levels in the brain.

individualized education program. A documented educational plan tailored to the needs of a given child, and mandated by the Individuals with Disabilities Education Act.

Individuals with Disabilities Education Act (IDEA). A U.S. law that ensures that students with a disability are provided with free and appropriate public education that is tailored to their individual needs.

infantile autism. An old term for autism intended to capture its early onset.

insertions. Mutations characterized by the gain of small or large segments of DNA.

intellectual disability. Deficits in the ability to reason, think, and learn that limit the development of the skills needed to lead a productive and independent life.

International Classification of Diseases (ICD). A diagnostic manual maintained by the World Health Organization.

introns. DNA information in genes interspersed with exons that does not code for amino acids.

karyotype. An image of the complement of chromosomes carried by an individual.

Light It Up Blue. An annual event on April 2, World Autism Awareness Day, during which landmark buildings and bridges throughout the world are illuminated with blue lights.

lipid. Nonpolar molecules, including fats, waxes, sterols, fat-soluble vitamins, glycerides, and phospholipids, that are commonly found in biological systems.

loss-of-function mutation. A mutation that results in partial or complete inactivation of the function of a gene.

magnetic resonance imaging (MRI). A clinical imaging technique that provides detailed anatomical pictures that can be used to diagnose diseases and injuries.

manic. A mood state during which people are energetic and happy.

median. The middle number in a sorted list of numbers.

messenger RNA (mRNA). An RNA that carries the protein-encoding information of exons and is competent for translation.

metaphorical language. The use of an object to represent an abstract idea.

Modified Checklist for Autism in Toddlers, Revised (M-CHAT-R). A diagnostic screening tool that asks parents or other caregivers a series of 20 questions about the behavior of a child and uses the answers to determine a low, medium, or high risk of ASD.

monogenic. When a disease or condition is caused by mutations in a single gene.

monogenic syndromic ASD. ASD in the context of a monogenic genetic syndrome.

monozygotic twins. Identical twins produced by a single fertilization event.

MSSNG. Pronounced "missing," a database of genomic DNA sequences from thousands of ASD patients and their family members.

mutagens. Physical or chemical agents that cause mutations.

mutations. Changes in the DNA sequences that might be deleterious, beneficial, or neutral.

neurofibromatosis type 1 (NF1). A monogenic syndrome that causes the growth of benign tumors on or just beneath the skin and along nerve fibers throughout the body.

neurons. The basic cellular units of the nervous system that carry nerve signals in the form of electrical impulses.

neurotransmitters. Chemicals that carry a nerve signal across the synapse between two neurons.

noncoding RNA. RNA produced by transcription of DNA that does not code for protein.

obsessive–compulsive disorder (OCD). Characterized by excessive thoughts that lead to repetitive behaviors.

oxytocin. A naturally occurring hormone that functions as a neurotransmitter and plays a role in sexual reproduction, childbirth, lactation, relationship formation, and social interaction.

pedantic speech. An atypical manner of speech during which children use formal language to try to speak with authority and sound like an adult.

perseverative speech. The repetition of a word or phrase after the stimulus that led to it has stopped.

personalized medicine. The use of genomic information to tailor a course of treatment not just to a particular disease, but to a particular patient.

pervasive developmental disorder not otherwise specified (PDD-NOS). A developmental disorder characterized by severe and pervasive impairment of reciprocal social interaction or communication skills and stereotyped behaviors, interests, and activities.

phenotype. A trait that results from a genotype.

pica. An eating disorder characterized by an unusual appetite for things that do not have any nutritional value, such as soil, rocks, ice, glass, hair, or feces.

picture exchange communication system. An ASD therapy during which a child is encouraged to initiate communication by the offering of a picture in exchange for something that the child wants, such as a toy.

pivotal response treatment (PRT). A therapy during which a trained professional uses reinforcements and motivations to encourage improvement in communication and social behavior.

plasticity. The ability of the brain to reorganize itself throughout life with new neuronal connections. Also called neuroplasticity.

polymerase chain reaction (PCR). A molecular genetics method that is widely used for the amplification of DNA with applications in basic research, forensics, and medicine.

postsynaptic. Refers to a neuron that receives neurotransmitters from a synapse during transmission of an impulse.

Prader–Willi syndrome. A neurological genetic disease caused by a paternally inherited deletion of part of chromosome 15.

presynaptic. Refers to a neuron that releases neurotransmitters into a synapse during transmission of an impulse.

pronoun reversal. A speech pattern in which children refer to themselves in the second person instead of the first person.

Punnett square. Used to illustrate possible offspring genotypes and their frequencies of occurrence.

recessive. An allele that does not produce a phenotype when present with a dominant allele in a heterozygous individual.

restricted and repetitive behaviors (RRBs). Repetitive movements and ritualistic behaviors displayed by children and adults with ASD.

rhosin. An inhibitor of RhoA, a GTPase that is required for the formation of synapses.

risperidone. A drug approved by the U.S. FDA in 2006 for the treatment of severe irritability displayed by children with ASD who are 5 years or older.

RNA splicing. Removal of introns and splicing together of successive exons to make an mRNA that contains a contiguous protein-encoding sequence.

savant syndrome. Characterized by prodigious talents or intellectual abilities in the background of deficits in other aspects of cognition that cause functional disability.

Screening Tool for Autism in Toddlers and Young Children (STAT). Assesses the risk of ASD in children between the ages of 24 and 36 months.

selective serotonin reuptake inhibitor (SSRI). A drug that limits the reabsorption of the neurotransmitter, serotonin.

sensory integration (SI). A therapy that uses structured play activities to change the way that a child reacts to sights, sounds, textures, and tastes.

sensory processing disorder (SPD). Characterized either by hypersensitivity to stimuli, extreme responses to loud noises, and a dislike for being touched, or by hyposensitivity to stimuli, a high pain tolerance, and a constant need to touch people and things.

Simons Simplex Collection (SSC). A group of over 2,500 families that include someone who has been diagnosed with ASD and that has participated in several large-scale DNA sequencing analyses.

single nucleotide polymorphisms (SNPs). Approximately 10 million places in the human genome where the DNA base present in one person is different in another person.

Smith–Lemli–Opitz syndrome (SLOS). A monogenic syndrome that causes microcephaly, low-set-ears, cleft lip or palate, extra fingers or toes, sleep disorders, atypical reactions to sensory stimuli, and intellectual disability.

spontaneous mutation. The conversion of a normal allele into a pathogenic allele during the process of sperm cell or egg cell production.

stereotyped speech. Persistent repetitive speech that occurs as a symptom ASD.

subclinical seizures. Seizures that do not produce observable symptoms and can only be detected as abnormal brain electrical activity during an EEG.

suramin. A drug used to treat parasitic diseases such as African sleeping sickness and river blindness that was shown to correct behavioral and neuropathological abnormalities in a mouse model of ASD.

training and education of autistic and related communication handicapped children (TEACCH). An educational program that uses structured teaching in a classroom setting to provide an environment in which children with ASD can learn.

transcription. The process by which the information in the form of the nucleotide sequence of DNA is converted into the nucleotide sequence of RNA.

translation. The process by which the nucleotide sequence of mRNA is used to direct the synthesis of proteins.

tuberous sclerosis. A monogenic syndrome that causes benign tumors to form in the organs throughout the body such as the brain, kidneys, liver, lungs, skin, and eyes.

tumor suppressor. A gene that protects a cell from becoming a tumor cell.

twin studies. Studies that use comparisons between identical twins and fraternal twins to separate out the effects of genetics and environment.

valproate. An epilepsy and bipolar medication that increase the occurrence of ASD when used during pregnancy.

Vineland Adaptive Behavior Scale. Assesses the ability of children to cope with environmental changes, to learn new skills for daily living, and to display independence.

Wechsler Adult Intelligence Scale. A widely used intelligence test for adults that yields an intelligence quotient (IQ).

Wechsler Intelligence Scale for Children. A widely used intelligence test for children that yields an IQ.

whole genome sequencing. DNA sequence of the protein-encoding exons of the human genome.

World Autism Day. Established as April 2 by the United Nations in 2007 to promote worldwide awareness of ASD.

zygote. A cell that results from the fertilization of an egg cell with a sperm cell.

Bibliography

Arking, D.E., D.J. Cutler, C.W. Brune, T.M. Teslovich, K. West, M. Ikeda, A. Rea, M. Guy, S. Lin, E.H. Cook, and A. Chakravarti. 2008. "A Common Genetic Variant in the Neurexin Superfamily Member CNTNAP2 Increases Familial Risk of Autism." *American Journal of Human Genetics* 82, no. 1, pp. 160-164.

Asperger H. 1944. "Die Autistisehen Psychopathen im Kindesalter." *Archiv für Psychiatrie und Nervenkrankheiten* 117, pp. 76-136.

"AutDB." Autism Informatics Portal. http://autism.mindspec.org/autdb/Welcome.do (accessed October 24, 2017).

"Autism Society." Autism Society. http://www.autism-society.org/ (accessed September 20, 2017).

"Autism Speaks." Autism Speaks. https://www.autismspeaks.org/ (accessed September 20, 2017).

Autism Spectrum Disorders Working Group of the Psychiatric Genomics Consortium. 2017. "Meta-Analysis of GWAS of over 16,000 Individuals with Autism Spectrum Disorder Highlights a Novel Locus at 10q24.32 and a Significant Overlap with Schizophrenia." *Molecular Autism* 8, no. 1, pp. 21. doi:10.1186/s13229-017-0137-9

Bacon, C., M. Schneider, C. Le Magueresse, H. Froehlich, C. Sticht, C. Gluch, H. Monyer, and G.A. Rappold. 2015. "Brain-Specific Foxp1 Deletion Impairs Neuronal Development and Causes Autistic-like Behaviour." *Molecular Psychiatry* 20, no. 5, pp. 632-639.

Basu, S.N., R. Kollu, and S. Banerjee-Basu. 2009. "AutDB: A Gene Reference Resource for Autism Research." *Nucleic Acids Research* 37, no. 1, pp. D832-D836.

Ciaccio, C., L. Fontana, D.A. Milani, S. Tabano, M. Miozzo, and S. Esposito. 2017. "Fragile X Syndrome: A Review of Clinical and Molecular Diagnoses." *Italian Journal of Pediatrics* 43, p. 39.

Dawson, G., J.M. Sun, K.S. Davlantis, M. Murias, L. Franz, J. Troy, R. Simmons, M. Sabatos-DeVito, R. Durham, and J. Kurtzberg.

2017. "Autologous Cord Blood Infusions Are Safe and Feasible in Young Children with Autism Spectrum Disorder: Results of a Single-Center Phase I Open-Label Trial." *STEM CELLS Translational Medicine* 6, no. 5, pp. 1332-1339.

DiStefano, C., A. Gulsrud, S. Huberty, C. Kasari, E. Cook, L.T. Reiter, R. Thibert, and S.S. Jeste. 2016. "Identification of a Distinct Developmental and Behavioral Profile in Children with Dup15q Syndrome." *Journal of Neurodevelopmental Disorders* 8, pp. 19.

"DNA Microarray Methodology." NC Community Colleges. https://www.youtube.com/watch?v=0ATUjAxNf6U (accessed December 23, 2017).

El-Fishawy, P., and M.W. State. 2010. "The Genetics of Autism: Key Issues, Recent Findings and Clinical Implications." *The Psychiatric Clinics of North America* 33, no. 1, pp. 83-105.

Escamilla, C.O., I. Filonova, A.K. Walker, Z.X. Xuan, R. Holehonnur, F. Espinosa, S. Liu, S.B. Thyme, I.A. Lopez-Garcia, D.B. Mendoza, N. Usui, J. Ellegood, A.J. Eisch, G. Konopka, J.P. Lerch, A.F. Schier, H.E. Speed, and C.M. Powell. 2017. "*Kctd13* Deletion Reduces Synaptic Transmission via Increased RhoA." *Nature* 551, no. 7679, pp. 227.

Feng, X., G.C. Ippolito, L. Tian, K. Wiehagen, S. Oh, A. Sambandam, J. Willen, R.M. Bunte, S.D. Maika, J.V. Harriss, A.J. Caton, A. Bhandoola, P.W. Tucker, and H. Hu. 2010. "Foxp1 Is an Essential Transcriptional Regulator for the Generation of Quiescent Naive T Cells during Thymocyte Development." *Blood* 115, no. 3, pp. 510-518.

Fernell, E., M. Anders Eriksson, and C. Gillberg. 2013. "Early Diagnosis of Autism and Impact on Prognosis: A Narrative Review." *Clinical Epidemiology* 5, pp. 33-43.

Garg, S., J. Green, K. Leadbitter, R. Emsley, A. Lehtonen, D. Gareth Evans, and S.M. Huson. 2013. "Neurofibromatosis Type 1 and Autism Spectrum Disorder." *Pediatrics* 132, no. 6, pp. e1642-e1648.

Howsmon, D.P., Uwe Kruger, S. Melnyk, S. Jill James, and J. Hahn. 2017. "Classification and Adaptive Behavior Prediction of Children with Autism Spectrum Disorder Based upon Multivariate Data Analysis of Markers of Oxidative Stress and DNA Methylation." *PLOS Computational Biology* 13, no. 3, pp. e1005385.

Jamain, S., H. Quach, C. Betancur, M. Rastam, C. Colineaux, I. Carina Gillberg, H. Soderstrom, B. Giros, M. Leboyer, C. Gillberg, T. Bourgeron. 2003. "Mutations of the X-Linked Genes Encoding Neuroligins NLGN3 and NLGN4 Are Associated with Autism." *Nature Genetics* 34, no. 1, pp. 27-29.

Kanner L. 1943. "Autistic Disturbances of Affective Contact." *Nervous Child* 2, pp. 217-250.

LeClerc, S., and D. Easley. 2015. "Pharmacological Therapies for Autism Spectrum Disorder. A Review." *Pharmacy and Therapeutics* 40, no. 6, pp. 389-397.

Li, D., H.-O. Karnath, and X. Xu. 2017. "Candidate Biomarkers in Children with Autism Spectrum Disorder: A Review of MRI Studies." *Neuroscience Bulletin* 33, no. 2, pp. 219-237.

Ma, D.Q., D. Salyakina, J.M. Jaworski, I. Konidari, P.L. Whitehead, A.N. Andersen, J.D. Hoffman, S.H. Slifer, D.J. Hedges, H.N. Cukier, A.J. Griswold, J.L. McCauley, G.W. Beecham, H.H. Wright, R.K, Abramson, E.R. Martin, J.P. Hussman, J.R. Gilbert, M.L. Cuccaro, J.L. Haines, and M.A. Pericak-Vance. 2009. "A Genome-Wide Association Study of Autism Reveals a Common Novel Risk Locus at 5p14.1." *Annals of Human Genetics* 73, no. 3, pp. 263-273.

"M-CHAT." Modified Checklist for Autism in Toddler (M-CHAT). https://www.m-chat.org/ (accessed October 18, 2017).

Menashe, I., E.C. Larsen, and S. Banerjee-Basu. 2013. "Prioritization of Copy Number Variation Loci Associated with Autism from AutDB—An Integrative Multi-Study Genetic Database." *PloS One* 8, pp. e66707.

"National Autism Association." National Autism Association. http://nationalautismassociation.org/ (accessed September 22, 2017).

Onay, H., D. Kacamak, A.N. Kavasoglu, B. Akgun, M. Yalcinli, S. Kose, and B. Ozbaran. 2016. "Mutation Analysis of the NRXN1 Gene in Autism Spectrum Disorders." *Balkan Journal of Medical Genetics* 19, no. 2, pp. 17.

Reilly, J., L. Gallagher, J.L. Chen, G. Leader, and S. Shen. 2017. "Bio-collections in Autism Research." *Molecular Autism* 8, p. 34.

Tabuchi, K., J. Blundell, M.R. Etherton, R.t E. Hammer, X. Liu, C.M. Powell, and T.C. Südhof. 2007. "A Neuroligin-3 Mutation

Implicated in Autism Increases Inhibitory Synaptic Transmission in Mice." *Science (New York, N.Y.)* 318, no. 5847, pp. 71-76.

"The National Autistic Society," The National Autistic Society. http://www.autism.org.uk/ (accessed October 11, 2017).

Thomas, A.M., M.D. Schwartz, M.D. Saxe, and T.S. Kilduff. 2017. "Cntnap2 Knockout Rats and Mice Exhibit Epileptiform Activity and Abnormal Sleep-Wake Physiology." *Sleep* 40, no. 1.

Thurm, A., E. Tierney, C. Farmer, P. Albert, L. Joseph, S. Swedo, S. Bianconi, I. Bukelis, C. Wheeler, G. Sarphare, D. Lanham, C.A. Wassif, and F.D. Porter. 2016. "Development, Behavior, and Biomarker Characterization of Smith-Lemli-Opitz Syndrome: An Update." *Journal of Neurodevelopmental Disorders* 8, p. 12.

Tick, B., P. Bolton, F. Happe, M. Rutter, and F. Rijsdijk. 2016. "Heritability of Autism Spectrum Disorders: A Meta-analysis of Twin Studies." *Journal of Child Psychology and Psychiatry, and Allied Disciplines* 57, no. 5, pp. 585-595.

Treffert, D.A. 2009. "The Savant Syndrome: An Extraordinary Condition. A Synopsis: Past, Present, Future." *Philosophical Transactions of the Royal Society B: Biological Sciences* 364, no. 1522, pp. 1351-1357.

Vignoli, A., F. La Briola, A. Peron, K. Turner, C. Vannicola, M. Saccani, E. Magnaghi, G.F. Scornavacca, and M.P. Canevini. 2015. "Autism Spectrum Disorder in Tuberous Sclerosis Complex: Searching for Risk Markers." *Orphanet Journal of Rare Diseases* 10, p. 154.

Weiss, L.A., and D.E. Arking. 2009. "A Genome-Wide Linkage and Association Scan Reveals Novel Loci for Autism." *Nature* 461, no. 7265, pp. 802-808.

Yoo, H. 2015. "Genetics of Autism Spectrum Disorder: Current Status and Possible Clinical Applications." *Experimental Neurobiology* 24, no. 4, pp. 257-272.

About the Author

Dr. Todd T. Eckdahl earned a BS in chemistry from the University of Minnesota, Duluth, and a PhD in molecular genetics from Purdue University. He is a professor of biology at Missouri Western State University, where he teaches genetics and conducts research with undergraduate students with support from the National Science Foundation. Dr. Eckdahl has published over 40 articles on molecular genetics, synthetic biology, and undergraduate science education. He is a member of the Missouri Academy of Science, the Genome Consortium for Active Teaching, and the Genomics Education Partnership. Dr. Eckdahl has been recognized for his teaching and research with the Missouri Governor's Award for Excellence in Teaching, the Missouri Western Board of Governors Distinguished Professor Award, the James V. Mehl Outstanding Faculty Scholarship Award, the Missouri Western Alumni Association Distinguished Faculty Award, and the Jesse Lee Meyers Excellence in Teaching Award.

Index

OTHER TITLES IN OUR HUMAN DISEASES AND CONDITIONS COLLECTION

A. Malcolm Campbell, *Editor*

- *Hereditary Blindness and Deafness: The Race for Sight and Sound* by Todd T. Eckdahl
- *Hemophilia: The Royal Disease* by Todd T. Eckdahl
- *Sickle Cell Disease: The Evil Spirit of Misshapen Hemoglobin* by Todd T. Eckdahl
- *Auto-Immunity Attacks the Body* by Mary E. Miller
- *Huntington's Disease: The Singer Must Dance* by Todd T. Eckahl
- *Nerve Disease ALS and Gradual Loss of Muscle Function: Amytrophic Lateral Sclerosis* by Mary E. Miller
- *Infectious Human Diseases* by Mary E. Miller
- *Breast Cancer: Medical Treatment, Side Effects, and Complementary Therapies* by K.V. Ramani, Hemalatha Ramani, B.S. Ajaikumar, and Riri G. Trivedi
- *Acquired Immunodeficiency Syndrome (AIDS) Caused by HIV* by Mary E. Miller
- *Down Syndrome: The Amazing Cookie* by Todd T. Eckdahl
- *Diseases Spread by Insects or Ticks* by Mary E. Miller

Momentum Press offers over 30 collections including Aerospace, Biomedical, Civil, Environmental, Nanomaterials, Geotechnical, and many others. We are a leading book publisher in the field of engineering, mathematics, health, and applied sciences.

Announcing Digital Content Crafted by Librarians

Concise e-books business students need for classroom and research

Momentum Press offers digital content as authoritative treatments of advanced engineering topics by leaders in their field. Hosted on ebrary, MP provides practitioners, researchers, faculty, and students in engineering, science, and industry with innovative electronic content in sensors and controls engineering, advanced energy engineering, manufacturing, and materials science.

Momentum Press offers library-friendly terms:
- *perpetual access for a one-time fee*
- *no subscriptions or access fees required*
- *unlimited concurrent usage permitted*
- *downloadable PDFs provided*
- *free MARC records included*
- *free trials*

The **Momentum Press** digital library is very affordable, with no obligation to buy in future years.

For more information, please visit **www.momentumpress.net/library** or to set up a trial in the US, please contact **mpsales@globalepress.com**.

CPSIA information can be obtained
at www.ICGtesting.com
Printed in the USA
LVHW050849171221
706343LV00004B/57